READ THE REVIEWS

"*Great coverage—**Project Workflow Management** takes the processes of project management to a new, higher level of competency. It addresses all essential project management processes needed to ensure a successful project outcome. A plethora of great examples and detailed guidance will complement the advanced skills of seasoned managers, as well as developing the skills of managers leading their first project. The effective planning, executing, and tracking process guidance prescribed for each 'frame' will serve readers and their organizations well in keeping projects on schedule and on budget!*"

—**Gerrard M. Hill**, Principal Consultant, Hill Consulting Group,
Author, *The Complete Project Management Office Handbook* and
The Complete Project Management Methodology and Toolkit

"*Wow! **Project Workflow Management** will help you take your ability to manage projects to the next level. It is a complete, easy to understand reference guide for ensuring strong project management processes in your organization. Use this book on your next project—regardless of its size—and you'll find yourself less frustrated and stressed out—and your projects far more successful!*"

—**Gina Abudi**, President, Abudi Consulting Group, LLC,
Co-Author, *Project Pain Reliever* and
The Complete Idiot's Guide to Best Practices for Small Business

"***Project Workflow Management** is a new approach to project execution. This book includes detailed descriptions of processes with examples where appropriate. The supporting diagrams and tables make it possible for you to adopt this approach on your projects, even with very little other formal project management training. It would also be useful at a corporate level, especially for companies looking to formalize project management processes and methods within their teams. Full of templates, this is a comprehensive resource that walks you through the processes with detailed flow diagrams and clear guidance for making your projects a success.*"

—**Elizabeth Harrin**, Director, The OTOBOS Group, Multiple awards winner,
Author, *Project Manager in the Real World* and *Social Media for Project Managers*

"***Project Workflow Management** is an ambitious book, with the potential to take project management in a new direction. We will have a lot more project managers in the world, by viewing project management through this lens. Rich Maltzman and Dan Epstein present a novel and creative way to look at the 'how' and 'when' of project management. The concept of framing project management as a process workflow makes the field accessible and easy to learn by almost anyone. It's like having a map in your hand and tracing where you want to go. The process workflow diagrams, decision points, templates and guidance in this book are excellent pointers. I applaud the authors for challenging our entrenched beliefs in the current project management methodologies.*"

—**Samir Penkar**, Founder, Future of Project Management,
Author, *A Project Management Journey*

"***Project Workflow Management*** *condenses decades of hard-won experience into a recipe for successfully completing projects of enormous complexity. While knowing what to do is a great start, the book guides you expertly through exactly how to do it. The wisdom in this reference will save you and your team from dozens of predictable and avoidable 'surprises' that you are bound to encounter—enabling you to benefit from the nuts and bolts approaches shared in this comprehensive guide to achieving outstanding results.*"

—**Kimberley Wiefling**, Founder & President of Wiefling Consulting,
Author, *Scrappy Project Management* and *Scrappy Women in Business*

"*Reading* ***Project Workflow Management*** *is the equivalent of a college course and spending two years managing projects inside a highly-functioning organization. It provides new project managers with a clear and tactical process flow for their responsibilities, and arms experienced project managers with a way to communicate the frame-by-frame project flow and the responsibilities for the project manager, delivery team, various management roles, and the sponsor. This is a book you will share with employees and mentees.*"

—**Kay Wais**, Founder and Principal of Successful Projects,
Vice President of AVS Group

"***Project Workflow Management*** *is an excellent guide for anyone starting to study, teach or work on project management. It provides a comprehensive list of what needs to be done in a project, while explaining how it can be done. The richness of unique concepts, formulas and templates in this book makes it a one-stop reference for all experienced project managers and for organizations that are standardizing their project management methodologies.*"

—**Jean Binder**, Senior Project Manager,
Author, PMI Cleland Award winner for *Global Project Management*

"***Project Workflow Management*** *by Dan Epstein and Rich Maltzman is a comprehensive guide to project management process workflows. It provides detailed explanations of each project management process and how it fits into the overall project life cycle. This methodical book will greatly help any project manager navigate and control a project.*"

—**Susan Madsen**, Project Management Leadership Coach,
Author, *The Project Management Coaching Workbook—Six Steps to Unleashing Your Potential*

"*Maltzman and Epstein provide a comprehensive, yet easy to understand and apply business process management approach to project management workflow that will help any organization improve their project management DNA. Great read and reference for project managers, PMO managers, and all those involved in project-related work.*"

—**Mark Price Perry**, Senior VP and Founder of BOT International,
Author, *Business Driven PMO Book Series*

"***Project Workflow Management*** *is an invaluable guide for both experienced project managers and 'newbies.' It's ideal for those of us who want to start with the 'big picture' and work our way down to the thousand little details that can make or break our projects…and our sanity.*"

—**Wayne Turmel**, President of GreatWebMeetings.com,
Host of "The Cranky Middle Manager Show," Chicago, IL

Foreword by HAROLD KERZNER, Ph.D.

PROJECT WORKFLOW MANAGEMENT

A BUSINESS PROCESS APPROACH

DANIEL EPSTEIN
RICH MALTZMAN

ISBN 978-1-60427-092-1

Printed and bound in the U.S.A. Printed on acid-free paper.
10 9 8 7 6 5 4 3 2 1

Library of Congress Cataloging-in-Publication Data

Epstein, Dan, 1943-
Project workflow management : a business process approach / Dan Epstein and Rich Maltzman.
pages cm
Includes index.
ISBN 978-1-60427-092-1 (hardcover : alk. paper) 1. Project management. 2. Workflow. I. Maltzman, Richard. II. Title.
HD69.P75E67 2013
658.4′04—dc23

2013033268

Direct all inquiries to J. Ross Publishing, Inc., 300 S. Pine Island Road, Suite #305, Plantation, Florida 33324.

Phone: (954) 727-9333
Fax: (561) 892-0700
Web: www.jrosspub.com

Table of Contents

Foreword

by Dr. Harold Kerzner
Senior Executive Director for Project Management
The International Institute for Learning

Project Workflow Management: A Business Process Approach is not the ultimate panacea to cure all of your project management ills. But from where we stand right now, it appears to be the best alternative, and significantly more valuable to the project manager than complex methodologies and processes.

When project management first began to evolve, corporate America was faced with the challenge of trying to determine whether an individual who desired to become a project manager had the potential to become a good project manager. We prepared a series of questions and asked them during an interview process, and sometimes we would ask the individual to attend an assessment center where he or she would be placed under stressful situations to see how the candidate would react.

While all of these approaches had merit, they soon failed. We discovered that people with Ph.D. degrees did not necessarily make good project managers. People with a command of technology often made worse project managers than people who just had an understanding of technology. At one point, we even demanded that only people with superior writing skills could become project managers, and that approach also failed. Simply stated, we were not sure what skills project managers needed to be successful.

Given our inability to clearly define the skills needed to be successful as a project manager, combined with the fact that every project has its own unique set of characteristics and every company has its own unique culture for managing projects, we began to focus our attention more so on methodologies and processes rather than on individual skills. We created methodologies that were "rigid" approaches, which was similar to placing handcuffs on the project managers. Methodologies were broken down into life cycle phases, and within each phase were processes based upon policies and procedures that, once again, reduced the flexibility that the project managers needed to meet the clients' requirements. Methodologies thrived on standardization and consistency. The typical methodology was seen by the project manager as an inflexible constraint.

Perhaps the primary reason why methodologies contained very little flexibility was because management did not trust the project managers to always make the right decisions. Guidance was needed, and it was provided through the methodology. But as project management began to mature and management began trusting the project managers to make good project decisions, methodologies and processes were replaced with forms, guidelines, checklists and templates. Methodologies were replaced with frameworks. It was like a cafeteria. The project manager would enter the cafeteria and take whatever forms, guidelines, templates and checklists off of the cafeteria's shelves that he or she believed were needed to provide the deliverables to the client.

While this approach appeared to have marginal success, project management was about to evolve even further:

- Project management was now being applied to more complex projects, such as projects related to the execution and implementation of strategic planning activities. These projects were previously managed by functional managers but were now assigned to project managers. Traditional methodologies and processes were not easily applied to these projects.
- These complex projects required an understanding of workflow across the entire company rather than just affecting a few functional areas.
- The project objectives were moving targets rather than stationary targets and could not be effectively managed with traditional project management methodologies and processes. *Workflow knowledge was necessary.*
- The more complex projects required that project managers have an understanding of the business as well as an understanding of the project.
- Project managers were now expected to make business decisions as well as project decisions.
- Project management methodologies now included business processes as well as pure project management processes.
- The traditional triple constraints on a project were replaced by competing constraints, many of which were dictated by business factors that emanated from the enterprise environmental factors.
- Project managers were provided with several tools to use while managing the project. Many of these tools were new to the project managers, and the project managers were unsure as to how and when to apply these tools.
- Project managers now viewed themselves as managing part of a business rather than simply managing a project.

We now began discovering flaws in our definition of project management. Historically, project management was defined as a series of decisions that had to be made at the right time and by the right people.

Methodologies, processes, forms, guidelines, templates and checklists were helpful but not sufficient by themselves. What is needed is a comprehensive understanding—and following—of workflow that shows the interaction of all participants and decisions needed at certain workflow points. Without a comprehensive understanding of project workflow, project decision making is a shot-in-the-dark or best-guess approach.

Without workflow knowledge, you can expect several of the laws of project management to exist in your project, such as:

- **Cohn's Law**—The more time you spend in reporting on what you are doing, the less time you have to do anything. Stability is achieved when you spend all of your time doing nothing but reporting on the nothing you are doing.
- **Cooke's Law**—In any decisive decision, the amount of relevant information available is inversely proportional to the importance of the decision.
- **Donsen's Law**—The specialist learns more and more about less and less until, finally, he knows everything about nothing, whereas the generalist learns less and less about more and more until, finally, he knows nothing about everything.
- **Dude's Law of Duality**—Of two possible events, only the undesirable one will occur.
- **Economist's Law**—If, on an actuarial basis, there is a 50-50 chance that something will go wrong, it will actually go wrong nine times out of ten.
- **Nonreciprocal Laws of Expectations**—Negative expectations yield negative results. Positive expectations yield negative results.
- **Fyffe's Axiom**—The problem-solving process will always break down at the point at which it is possible to determine who caused the problem.
- **Gresham's Law**—Trivial matters are handled promptly; important matters are never resolved.
- **Johnson's First Law**—When any mechanical contrivance fails, it will do so at the most inconveniently possible time.
- **Murphy's Seventh Law**—Left to themselves, things tend to go from bad to worse.
- **Pudder's Law**—Anything that begins well ends badly. Anything that begins badly ends worse.
- **Putt's Law**—Technology is dominated by two types of people—those who understand what they do not manage and those who manage what they do not understand.

Good methodologies "may" be able to prevent these laws from being applicable to your projects. But it is highly more likely that a good understanding of and adherence to project workflow can prevent most of them from happening.

Introduction

To the Reader

Since you are reading this, we assume that you are a project manager or perhaps you are considering venturing into the project management profession. You may also be a project management software developer, seeking to provide benefits to your clients by reducing the cost of projects. Project management is indeed a very exciting and rewarding profession, but at the same time, it is one of the most difficult jobs, often misunderstood by project team members and management alike. The project manager will never win a popularity contest, because even though he or she is not usually a personnel manager of team members, he or she nevertheless sets work deadlines, demanding status reporting and adherence to the project work rules and schedule. These demands—coming from someone who is not technically the team members' supervisor—will not necessarily win you any favors with team contributors. Most project managers experience tight budgets, sometimes unrealistic deadlines set by management and a chronic lack of qualified human resources.

Luckily, today, industries largely understand the value of project management and treat the project management discipline with its deserved level of respect. Still, there are many extremely frustrating circumstances which may push the less experienced project manager to question the wisdom of getting involved with this profession. Here are some examples of such frustrating situations drawn directly from the authors' experience:

- The functional manager, who directly controls some of the resources assigned to your project team, demands that you do not speak to his employees directly. Before you assign any task to an employee, he says, you must submit task details in writing to him. If the task is approved, it is sent to the employee by the functional manager. It is impossible to accomplish anything in this work environment, let alone deliver the project on time within a tight budget, considering that even a medium-size project has hundreds of mutually dependent tasks.

- In the middle of the project, the line manager pulls out her resources and reassigns them to another project, leaving you out in the cold.
- Your own boss, following a "gut feeling," promises the client delivery of a part of your project within a month, while the carefully developed schedule indicates a true period of three months. No additional resources or budget are available.
- Some of your team members believe that the project manager is merely an administrator, serving the basic function of a "team secretary." They grudgingly accept assignments from you, saying they will do it on their own schedule and let you know when it is done, the project schedule notwithstanding.
- Your business clients are in the habit of bringing new scope change requests directly to your team members in order to "reduce bureaucracy." You only become aware of this when you discover unplanned activities, resulting in schedule delay and extra costs, which are not covered by the project schedule and budget and are incompatible with other elements of the project.

These are (hopefully) exceptions in most environments, but may happen every day in some workplaces. Knowing how to recognize those work environments is essential to taking preventive steps to avoid many of these issues.

This book is written to alleviate some of the problems described above and make it easier for you to manage projects. The book clearly identifies roles and responsibilities of all project staff, including management, making their personal biases, guesses, incorrect assumptions and "gut feelings" irrelevant. The PM Workflow framework, which is the book's foundation, is like the project GPS. A GPS automatically selects the most efficient route to drive, rerouting a car when road construction or heavy traffic is encountered. Similarly, under the PM Workflow guidance, a project flow execution path is based on the system of requests, decision checkpoints, decision tables and project health evaluation results, directing the project flow and continuously prompting and instructing project managers. This unique approach simplifies daily tasks of managing projects and significantly improves project quality, because compliance with all required project management processes, which is a major factor for building quality into the end product, is enforced by PM Workflow.

Why This Book Is Different from All Other Project Management Books

Project Workflow Management: A Business Process Approach is the first book ever in the market based on project management process workflow, often referred to as PM Workflow. This book describes a step-by-step workflow guiding approach, which significantly simplifies project management activities by prompting execution of all required processes on time and redirecting to an alternative path in case of project issues. This is a unique approach which prepares readers to manage the *entire project life cycle.*

The book presents project management as a single (but rather complex) multithreaded business workflow that breaks down into a set of detailed processes with logical decision making, control point tests and loops. By definition, *workflow* is a means to identify and diagram procedural steps and logic used to achieve a specific goal. It shows the direction of movement, decision points, alternative processes and dependencies. It identifies the start and end points of each element of the workflow and often identifies who reviews the implementation and how it is done. One of the important elements of the workflow is evaluating the results of executing each process in the workflow. The workflow consists of the detailed executable processes.

Process is very different from a *process model.* Process models are presented in *A Guide to the Project Management Body of Knowledge* (*PMBOK® Guide*), Six Sigma and project management methodologies, even though they are often called processes by these entities. By definition, process models are descriptions of processes, outlining expectations of the process implementation. Process is the description of the physical operation for the process model implementation. In other words, *process* is really the *implementation of the process model design* in order to achieve specific goals. A step-by-step sequence of the elementary processes is a workflow. The elementary process definition is given in the following chapters of this book.

The execution of the PM Workflow will control the project flow and will prompt and instruct project managers in every step of the workflow—from day one to the project closing—while constantly performing qualitative assessments of outcomes and suggesting steps for resolution. After reading this book, even an inexperienced individual will be armed with sufficient knowledge of project workflow management and practical steps and tools to confidently manage a project from A to Z. Along with the workflow, the book describes many project management methods, tips and techniques, showing examples and providing readers with the practical knowledge required to manage projects hands-on from scratch, up to and including the delivery of the completed and tested product or service to the client. The book will also serve as a reference for guiding project managers in every step of a project workflow.

Today, there are several leading project management resources produced by various institutions. The most popular are:

- The framework of practices, techniques and tools developed by the Project Management Institute (PMI®) and described in its *PMBOK® Guide.* The *PMBOK® Guide* is not considered by PMI® to be a methodology, because the process models described therein are generally accepted good practices and not a practical guide to manage projects. The *PMBOK® Guide* was first published in 1996 and claims to apply to "most projects, most of the time." It gives careful consideration to project scope, schedule and cost and provides a set of high-level process models for development of detailed project management processes and workflows.
- The Six Sigma methodology was developed by Motorola in 1986. Six Sigma employs a set of quality management methods using statistical analysis to identify causes of defects and establish repeatable sets of processes. Statistically, the term *six sigma* is related to quality, where 99.99966% of resulting products are statistically free of defects. While this is not the practical outcome, the methodology sets a very high level of expectations. Six Sigma applies equally to projects as to business as usual activities, which include operations and manufacturing.
- PRINCE2, an acronym for **Pro**jects **In** **C**ontrolled **E**nvironments, version **2**, is a project management methodology. This methodology covers management, control and organization of projects. It was developed by the U.K. government and

is used in the U.K. and partly in other parts of Europe and Australia. PRINCE2 consists of seven themes and seven corresponding high-level process models, such as Business Case, Organization, Quality, Plan, Risk and Progress. This methodology is considered by many to be unsuitable for small projects, because the project environment overhead is too high.

- Method 123, also called MPMM, is an attempt to combine the *PMBOK® Guide and PRINCE2.* One advantage of the methodology is that it provides some of the actual physical processes, rather than process models. It contains such process attributes as templates, forms and checklists.
- SCRUM is a set of high-level guidelines for Agile development, mentioned below. It is used in environments where it is difficult to do detailed planning. SCRUM utilizes empirical process control, which is based on observation of results produced and client inputs, rather than the traditional planned project management.

The existence of several project management methodologies often comes in conflict with those environments where specialists in several competing methodologies are available. This often leads to a tense environment. This evokes the expression "when there are seven nannies for one child, the child is often left unattended." A similar situation can exist in those project environments where there are no strict project management processes and there is not a single underlying methodology.

Recently, a very first significant step to fix this situation was made by the International Organization for Standardization, which came up with ISO 21500. The new standard is a very high level of concepts and process models to streamline the project management best practices worldwide. ISO 21500 deals not only with projects, but also with business and organizational environments, strategic alignment and social and environmental impact in project selection. This standard is for both project managers and for enterprise managers.

This new standard provides the universal practices for managing all projects worldwide. While it is aligned with the *PMBOK® Guide,* it is not the same. It introduces 10 subject groups (Integration, Stakeholder, Scope, Resource, Time, Cost, Risk, Quality, Procurement and Communication), each having distinct process models applicable to each process group, such as Initiating, Planning, Controlling, Implementing and Closing.

The project workflow management in this book is compliant with ISO 21500. There are minor differences in terminology, such as using the word "frame" instead of "phase," used in both ISO 21500 and the *PMBOK® Guide.*

The difference between frames and phases is that *phases* imply a sequential execution. Execution of *frames* instead depends entirely on the workflow. The process flow starts with the requirements management processes in the *Requirements Frame* and ends with the project closeout set of sequential processes in the *Closing/Testing Frame.* Based on the control point test conditions, the process flow may branch to specific processes in *any* frame. The description of all project management process flows is provided in the following chapters in detail. These chapters include applicable techniques, methods, tools, templates and checklists.

Our PM Workflow includes the physical initiation processes placed in the *Requirements Frame.* Requirements processes are executed at the beginning of a project and every time a project scope change is initiated during the entire life cycle, while initiation processes are used only once at the project start. Thus, instead of an initiation phase, there is a *Requirements Frame* in our project workflow. Also, since a project change request may come while running user acceptance testing or even user training, the process flow will be directed back to the *Requirements Frame.* However, during the execution of the actual closing processes, there is no flow back to the requirements or initiation processes, just as described by ISO 21500. Our *Closing Frame,* as opposed to the closing phase, includes the user acceptance test set of physical processes and the user training process, which are not mentioned in ISO 21500.

In accordance with the approach presented here, the overall high-level project management process is decomposed into a hierarchical tree of detailed physical processes. Each process consists of specific steps in four project *frames* (*Requirements, Planning/High-Level Design, Construction/Tracking and Closing/Testing*). Each frame is a collection of related processes in the overall workflow. Our project workflow physical processes fit well into the ISO 21500 high-level framework.

The book is written for the following audiences:

1. **Academic students of project management**—It can serve as a primary project management textbook.
2. **Project managers and business or technical professionals** interested in project management—As project managers, we are continuously students of our discipline. This book may also be used as a reference for this audience.
3. **Software companies** developing traditional or Cloud project management applications—An application developed based on the workflow presented will automate and simplify project management tasks.

As in any workflow, project workflow has a single entry point (start of project) and a single exit point (end of project). The process flow during the project life cycle will alternate between project frames and individual processes within frames. Depending on the results of many control point tests, such as a new scope change, new issue, new risk, quality issues, budget overrun, schedule problems and so on, the process may loop to another process area or simultaneously start executing a parallel process thread in the same or other project frames, which, upon their completion, merge again to continue on from there. For example, while tracking implementation of one of the project tasks in the *Construction/Tracking Frame,* a parallel thread of producing the scope change requirements may start in the *Requirements Frame,* then continue on to execution of the scope change planning process in the *Planning Frame* and then start actual implementation and tracking of the scope change in the *Construction Frame.* If serious project issues are encountered along the way, the process flow will be directed to the Troubled Project Assessment process and from there to the Take Corrective Action process.

This approach will minimize the need for the consultancy of experienced (and more costly) project mentors, as it will safeguard the less experienced project managers against most project life cycle errors; in doing so, *application of the workflow will reduce the cost of company projects.* A medium-level project manager will be able to take charge of large projects, currently controlled by the senior project managers, since every step of a project will be controlled by the process, with all project issues reported. Similarly, junior project managers, project coordinators or students who have studied this book and practiced using any project scheduling tool may substitute for the medium-level project managers. The experienced project managers can thus be freed up for roles in program and portfolio management, overseeing larger *collections* of projects.

Furthermore, *project quality will be greatly improved,* since compliance with all project management processes is enforced by the workflow, saving companies significant expenses.

Today, there are several major software applications for project management. They are widely used and impressive in capability, but none of them covers the entire project life cycle and they count on whatever structure or flow is entered. Using those applications, a project manager is able to plan and track any part of the project. However, the specifics or even inclusion of some essential project processes in the plan (such as quality management, risk management and so on) are accounted for by the individual project manager and corporate guidelines. Furthermore, although project tracking with those tools will accurately reflect the actual versus expected project outcomes, it will not make a qualitative assessment of any issues that arise, nor will it direct the project manager to suggested steps required for managing an issue. The quality of the plan produced with those tools correlates directly with the project manager's experience and ability, which in turn sets a requirement for skilled and highly paid project managers to run even fairly straightforward projects.

A software application based on the workflow will automate project activities for managing the entire project life cycle from project request submission to project closure, while constantly performing qualitative assessments of outcomes and suggesting steps for resolution. The application may seamlessly incorporate existing project management tools, such as MS Project, Clarity, Primavera and so on. This will apply equally to both the traditional and the Cloud-based applications. The application would have a major practical implication for the project management field, ensuring significant savings on every project by the end users and corresponding profits to developers of such an application.

As companion material to the book, three software tools are provided, which are explained in the book. The tools may be downloaded free from the publisher (at www.jrosspub.com) or the authors (at www.pm-workflow.com). See the Web Added Value™ page in the front matter for more information.

Acknowledgments

Grateful acknowledgment is made to our publisher, Drew Gierman, who recognized the potential of this book and provided guidance, where project management book editors from the big-name publishing houses did not seem to even know the difference between the project workflow and critical path analysis. Credit also goes to Drew for bringing both authors together to work on this book, enriching the final outcome via a blend of their specific knowledge, experience and focus areas—a blend which we feel ultimately benefited this book.

Our deepest thanks and admiration to Sarah Shah, who, with an extensive education in journalism and a healthy and avid interest in English (she teaches English in Washington, D.C.) volunteered to edit this book to meet the standards of our publisher, raising it from the level of technical jargon to nearly poetry. Sarah's contribution was critical, considering that one of the authors until 1971 lived in Russia and spoke mostly Russian. Perhaps equally challenging, the second author hails from New England.

About the Authors

Dan Epstein combines over 25 years of experience in the project management field and the best practices area, working for several major Canadian and U.S. corporations, as well as 4 years teaching university students project management and several software engineering subjects. He received a master's degree in electrical engineering from the LITMO University in Leningrad (today St. Petersburg, Russia) in 1970, was certified as a Professional Engineer in 1983 by the Canadian Association of Professional Engineers–Ontario and earned a master's certificate in project management from George Washington University in 2000 and the Project Management Professional (PMP®) certification from the Project Management Institute (PMI®) in 2001.

Throughout his career, Dan managed multiple complex interdependent projects and programs, traveling extensively worldwide. He possesses multi-industry business analysis, process reengineering, best practices, professional training development and technical background in a wide array of technologies. In 2004 Dan was a keynote speaker and educator at the PMI-sponsored International Project Management Symposium in Central Asia. He published several articles in company internal publications. In the summer of 2008 he published "Methodology for Project Managers Education" in a university journal.

Dan first started development of the PM Workflow in 2003, and it was used in a project management training course. Later this early version of the methodology was used for teaching project management classes at universities in the 2003–2005 school years. Later on, working in the best practices area, the author entertained the idea of presenting project management as a single multithreaded business workflow. In 2007–2008 the idea was further refined when teaching the project management class at a university. In 2009–2011 Dan continued working full time on the PM Workflow, completing development in 2012.

Rich Maltzman, PMP, received a BSEE in 1978 from the University of Massachusetts, Amherst, and an MSIE with a focus on ergonomics from Purdue University in 1982. Since then he has also received master's certificates in business management from Wharton (University of Pennsylvania) and INSEAD (France). He started his professional career in 1978 as an engineer at a major telecom for which he still works, but now in its global PMO. His project management career started in 1988 as a Project Management Director and included a two-year assignment in The Netherlands, where he built a team of project managers for telecom deployments in EMEA. He has directed large projects (1996 Summer Olympics telecom) and helped coordinate the merger of two large corporate PMOs.

In his global PMO work, Rich has been responsible for the learning and professional advancement (career path, competency models, accreditation, credentialing, training) of about 2,500 project managers, and in that role he has helped build a project management community by bringing it together with an annual International Project Management Day Symposium, Project Team of the Year awards, bimonthly project management roundtables and ongoing discussions on dedicated corporate social media channels for project managers.

Rich also does a significant amount of project management consulting and teaching at Boston-area colleges and universities as well as medium and large businesses and has partnered with companies that develop PMP® Exam Prep materials to create study aids such as flash cards and to compose practice exam questions. Rich has recorded video as a project management subject matter expert for online project management training programs. He is the co-founder of EarthPM, LLC, a firm that focuses on sustainability thinking in project management, and in that role is co-author of the book *Green Project Management,* winner of the PMI® 2011 Cleland Award for literature. Rich just completed a two-year term as Vice President of Professional Development for the Mass Bay Chapter of PMI®. He has presented (on invitation) on the topic of green project management at Project Management South Africa and for the Government of Malaysia and PMI® Chapters in New England, Costa Rica, Florida and Missouri. Recently, as part of EarthPM, Rich contributed to the fifth edition of the *PMBOK® Guide* and the ISO 21500 standard and with co-author David Shirley, PMP, has written chapters for *The AMA Handbook of Project Management* by Dinsmore and Cabanis-Brewin and *Sustainability Integration for Effective Project Management* by Silvius and Tharp.

Free value-added materials available from the Download Resource Center at www.jrosspub.com

At J. Ross Publishing we are committed to providing today's professional with practical, hands-on tools that enhance the learning experience and give readers an opportunity to apply what they have learned. That is why we offer free ancillary materials available for download on this book and all participating Web Added Value™ publications. These online resources may include interactive versions of material that appears in the book or supplemental templates, worksheets, models, plans, case studies, proposals, spreadsheets and assessment tools, among other things. Whenever you see the WAV™ symbol in any of our publications, it means bonus materials accompany the book and are available from the Web Added Value™ Download Resource Center at www.jrosspub.com.

Downloads for *Project Workflow Management: A Business Process Approach* consist of:

- A project risk assessment tool
- An earned value analysis tool
- A client satisfaction survey tool
- Instructions on how to use the above downloadable tools

All tools are developed on an MS Excel platform. They are intended to save project managers time calculating project risk scores, doing the earned value analysis and quantifying the level of client satisfaction based on the client's answers to several questions. The tools have never been intended as full-fledged applications. One of the book's authors developed them for his own use and they have served him well since 1998, updating them from time to time.

All tools are templates, meaning that a new copy of the tool will be created every time you open the template, leaving the template intact. You may save results as a project document under a meaningful name and access it later when needed.

All tools are in *.xltm* format, which means they are macro-enabled templates. In order to run the tools you must allow Excel running macros. None of the tools contain any macros, but they use some VB controls and run code, which will not work if macros are disabled. When saving results, the tools will automatically propose saving as a *.xlsx* regular spreadsheet, rather than a macro-enabled *.xlsm* spreadsheet. Excel will issue a warning that you may lose some functionality if you do that. If the saved work will be used only as a document, rather than a tool, then go ahead and save it. Just remember that the saved copy may have limited functionality if used as a tool. In order to run a fresh copy of the tool, run a corresponding template.

Because of the embedded code and VB controls, none of the current known Excel-compatible applications will work on android or Apple tablets.

Section I. Project Management Overview

Introduction to Project Management and Workflow

Why Project Management?

The long project development cycle and large number of defects left in products after their release into production affect not only the direct development costs, but also the lost revenue or even the loss of the opportunity itself. The remedy to the above problem is to assign high priority to well-managed development and the ability to predict cost, delivery date and quality. The only way to ensure this is the application of the discipline called project management.

The great business thinker Tom Peters predicted about 25 years ago in his book *Thriving in Chaos* that 90% of companies are likely to be either eliminated or transformed into third-generation companies within 10 to 20 years. Those third-generation companies are characterized by involvement of the "knowledge workers" in company decisions on all levels and most company activities either turned into projects or were doomed to failure.

During tough economic times, the ratio of project cost to benefits naturally draws serious attention and has prompted many companies in all industries to rethink their approach to project management.

The tremendous growth of the project management profession is due to the realization that project management is able to provide quality and a predictable budget and schedule for new developments. These factors often determine who the most successful and competitive industry leaders will be. In order to ensure professionalism, the Project

Management Institute (PMI®) introduced professional certification of the Project Management Professional (PMP®) and several other specialized certifications. PMP has become the worldwide recognized and exclusive credential, with an exponential growth to more than *half a million* credentialed PMPs.

Basic Project Management Concepts

Before describing the project management workflow, it is necessary to have a basic idea about project management concepts. Let's say you have to assume project management responsibilities on a brand new project. You will have many questions: What do you have to do to start and run it? What has to be accomplished? How much is it going to cost and when will it be finished? Who is going to pay for this? What constitutes a project, and what constitutes my specific project? What exactly is the project manager's job and how does it differ from a general managerial job? Let's review those basic concepts.

What Is a Project?

A project is a time-limited business function which has multiple mutually dependent activities scheduled between the planned start and end dates, whose purpose is to create a new and unique solution.

Thus, in the construction business, building a house or factory would constitute a project. Even though the building may look identical to the one across the street, each one of them requires separate individual planning steps and they are not necessarily going to be the same. Development of new software is also a project, just as is performing modification of existing software modules. Creation of a medical center in an existing building or establishing a manufacturing assembly line is a project, too. All the above examples have a scheduled beginning, a scheduled end and have as their product, respectively, a new building, a new software product, a new medical service and a new manufacturing assembly line.

On the other hand, consider a brand new automotive engine assembly line with all the machinery and quality control equipment installed. The task is to start manufacturing car engines on a specific day and to continue doing that for the whole year. This is not a project even though the start and end days are scheduled, because the assembly line is a *continuous repeating operation* to assemble exactly *the same engine again and again.*

What Is Project Management?

Project deliverables may be highly technical in nature, using the latest in modern technology. Despite this, project management is not a technical function, but rather is a *business* function consisting of the project management business processes and the individual project manager's application of the knowledge, techniques, methods and tools he or she has acquired to navigate through the process of managing a project.

While understanding of the practice area (or technology) by a project manager is encouraged, being a detailed technical specialist is not required. However, being fully competent as a businessperson is *essential.* The days when the best technical specialist was automatically considered the best project manager are long gone.

Project Manager's Duties and Responsibilities

The project manager is in charge of producing project deliverables, planning and running his or her project using the skills available in the project team. One could assert that project management is, in essence, about getting project deliverables completed through the work of others. Under the project manager's control, the precise scope of the project has to be defined, project plans developed, the work monitored and so on. All these will be described in great detail in the remainder of this book. The following are some examples of the project manager's duties and responsibilities:

1. Documenting detailed up-to-date project information
2. Assisting clients in defining project requirements
3. Planning the project
4. Staffing the project team
5. Estimating project work and labor required
6. Controlling project risks
7. Communicating to project team members and leading them
8. Reporting to management and clients
9. Managing client—and all stakeholder (see the note below)—expectations
10. Communicating to all stakeholders and maintaining productive stakeholder relationships
11. Controlling project financials
12. Controlling project scope changes
13. Assuring project quality
14. Organizing meetings, reviews, test activities and so on
15. Transitioning project deliverables to steady-state operation
16. Closing the project and releasing project resources

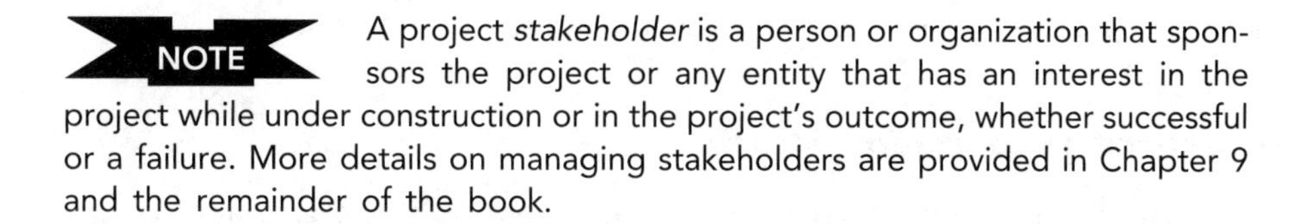

A project *stakeholder* is a person or organization that sponsors the project or any entity that has an interest in the project while under construction or in the project's outcome, whether successful or a failure. More details on managing stakeholders are provided in Chapter 9 and the remainder of the book.

Organization Structure

The power of the project manager to build the team and establish compliance with the project plan greatly depends on the organization's structure. There are three major types of organization structure, with some variations from those described below:

1. Functional
2. Matrix
3. Projectized

Functional Organization Structure

In the functional organization structure, the project manager has no authority over resources and cannot assign work to individuals. In this environment, project managers often are called project coordinators and usually have little or no decision or enforcement authority. Here, resources are managed by a functional manager according to the resources' specialization. In this environment you will find a director of manufacturing, director of mechanical engineering, director of electrical engineering, director of marketing and so on. Functional contributors (engineers and marketing specialists, for example) will likely report to line managers who may have different priorities than project managers, and therefore the project schedule may often change, making the specific project stakeholders' satisfaction elusive. The overall authority over projects is assigned to functional or line managers. Professional project managers are rarely happy working in this environment, because of the obvious lack of control and the barriers to good project communication.

Matrix Organization Structure

Here project managers have more power over resources than in the functional organization. Project managers in this environment report either to the same senior manager, just as line managers do (strong matrix), or to the manager of project managers (weak matrix). Line managers are still in charge of resource pools, but after they assign resources to the project, they cannot pull them without the higher level managers' consent. The strong matrix provides the project manager with a stronger position to argue resource issues with line managers and also directly with the second-level manager. Project managers plan resource allocation in percentage points of the full load. Thus, a resource may be allocated for 100% of the time for the first six months, then 50% for another two months and then released altogether. This provides better resource utilization, but requires very thorough planning. The resource requirements changes in one project may lead to resource issues in other projects which share the same resources.

Projectized Organization Structure

In this type of organization there are no line managers. All the organization's activities are broken into large projects, each led by a project manager. Usually this type of organization is effective in heavy industries, such as aircraft, automotive, spacecraft, military and shipbuilding, where each project is worth hundreds of millions of dollars. It is also found in architecture, law and advertising firms, where major projects (a building, a large lawsuit, a major advertising campaign) are taken on as *projects* for a particular customer. Here project managers have exclusive authority over resources for the project duration and have power to enforce standards, regulations and schedule. Project managers, who often are called program managers, are executives and report to the company's top management. The major drawback here is poor resource utilization, since it is practically impossible to utilize each resource 100% for the duration of the project. In addition, since there are no line managers in this environment, project managers have to perform this additional time-consuming duty for the duration of the project.

Where Do Projects Come From?

Consider the following project initiation scenarios:

1. A project delivery organization wins a bid from an outside company to develop a new product or service. This means that the organization's proposal was selected due to its past experience in that type of project, proven quality or the most cost-effective solution. Sometimes an outside company may award a project to an organization without a bid due to its proven reputation for great results.
2. A project starts due to the specific need of an organization, such as a business need for a new manufacturing line or a new software program which supports the expanded business. In this case the requesting department should outline benefits to justify project expenses.
3. Market research may have determined that there is a market demand for a specific service, system or product, which is predicted to be profitable to the company's business.

Undertaking any project is a business risk for the organization. The bigger the project and the less experienced the team, the larger the risk. The risk may reach such an extent that the company may jeopardize large financial assets, reputation and its very future existence. Project risks will be discussed later, but it is reasonable to expect that a company will not assign an inexperienced project manager to a multimillion-dollar project. It is also unlikely that a new, untested project manager will be assigned to manage a project for an important business client or for development of a brand new marketable product. If such a project turned out to be unsuccessful, the company finances and reputation may suffer beyond repair. In the worst-case scenario, the company may be liable for damages and fines and may even be forced into bankruptcy.

In internal noncritical projects, financial losses may still occur due to an unsuccessful project, but it is less likely that the company reputation will be damaged and fines and lawsuits are not a threat.

There are small differences between the above three types of projects at the initial and closing parts of a project, but the benefit of the project management discipline is that generally the project workflow will not change.

Contract Types

Regardless of the source of projects, they will be governed by a contract with an external client or a statement of work/document of understanding for internal ones. In all documents the schedule of payments will determine when and how the delivery organization or department will be paid for the project work produced. Most probably, for internal projects within an organization, each department has budgets for projects, equipment and so on. This type of financial discipline promotes responsible internal project expenditures by all departments. There are three major contract types, with some variations of each type:

1. Fixed price contract
2. Cost-reimbursable contract
3. Time and materials contract

Fixed Price Contract

In a fixed price contract, regardless of unexpected events during the course of the project, the price will stay the same, unless specifically defined in the contract. As variations of a fixed price contract, the contract may include incentives for the timely delivery of the project or penalties for late delivery. External clients prefer a fixed price contract because it minimizes the buyer's risk, transferring it instead to the seller. Remember that risk can be positive or negative, so this also provides the seller the opportunity to reap a greater profit by keeping costs low (since the seller would retain the advantage of those lower costs).

Despite this possibility of retaining lower costs, this is the least preferable contract type for project delivery organizations, because often they have to legally commit to the project price and duration before they even have all requirements. Unless your team has plenty of experience developing very similar projects, your risk will be very high. Many large companies that develop projects for external clients avoid taking fixed price projects, at least until project requirements analysis is complete.

Cost-Reimbursable Contract

Cost-reimbursable contracts involve the buyer paying the actual cost to complete the work. Of course, the seller needs to make a profit, so there are a variety of schemes for providing a fee or profit to the contractor, as well as incentives to prevent the seller from incurring unnecessary costs. Here, the buyer assumes the threat of cost increases and reaps the benefits of cost reductions.

Time and Materials Contract

In time and materials contracts, which have elements of both cost-reimbursable and fixed price contracts, rates are negotiated before the contract award; these rates take into account labor and materials. As work is completed, the contractor bills against the *rates* agreed to in the contract regardless of the actual cost. This in no way means that the project will have a bottomless budget and endless schedule. There are still legal commitments for delivering the project on time and within the estimated price, but the initial project cost estimates may be allowed to have a wider margin of error for the later frames. As more knowledge is gained about requirements and the project details, the accuracy of project estimates will improve.

NOTE For those interested in further information on contract types, we suggest that you visit the U.S. Defense Acquisition University site, which contains a wealth of knowledge on the subject (see https://acc.dau.mil/CommunityBrowser.aspx?id=24930 in particular).

Project Life Cycle

The project life cycle is the sequence of all project activities required in order to achieve project objectives.

NOTE This book attempts to keep terminology as close as possible in line with the *PMBOK® Guide*, but in some cases that is not entirely possible. One of the examples described in the Introduction is that the word *frame* is used here as a substitution for the word *phase*. The word phase implies sequential execution. Once a phase is over, the next phase in a sequence starts. Frames are not necessarily sequential. Processes in several frames may be executed at the same time. Each frame is a collection of related processes in the overall workflow.

The following project frames are established in this book:

- Requirements Frame
- Planning/High-Level Design Frame
- Construction/Tracking Frame
- Closing/Testing Frame

Throughout this book, the Planning/High-Level Design Frame often will be called the Planning Frame. The Construction/Tracking Frame will be called the Construction Frame, and the Closing/Testing Frame will be called the Closing Frame.

In the Requirements Frame, the major process goal is establishing justification for the project, allocating detailed project requirements and obtaining project funding. However, if a change request is generated in any frame during the project life cycle, the project flow loops back to the Requirements Frame in order to produce requirements for the scope change request. This frame is generally aligned with the Initiation process group of the *PMBOK® Guide*, but may have other processes related to requirements gathering, analysis, documenting and approval.

The goal of the Planning Frame is development of the detailed plan for the business solution and change requests, providing detailed project estimates and planning. The frame activities usually include the high-level design and architecture. There is additional use of processes in this frame, such as planning for risk containment, planning for major project issues, planning for project scope changes, etc. These all will be discussed later. This frame is generally aligned with the Planning process group of the *PMBOK® Guide*, but may have other processes related to planning and high-level design.

The goal of the Construction Frame is the development and implementation of the business solution and change requests, simultaneously monitoring and managing project risks, cost, schedule and quality of the solution. This frame will be used for tracking of the requirements process in the Requirements Frame, tracking of user acceptance testing in the Closing Frame, etc. This frame is generally aligned with the Monitoring and Controlling process group of the *PMBOK® Guide,* but may have other processes related to project health control.

The goal of the Closing Frame is verification that the solution satisfies the stated purpose of the project (user acceptance test), transition of the completed solution to the client, closing the project and releasing project resources. Also, some of this frame's processes may be initiated directly from any of the above frames when project termination is requested due to funding withdrawal, outdated requirements, etc. This frame is generally aligned with the Closing process group of the *PMBOK® Guide,* but may have other processes related to testing and user training.

In the last few years new methodologies for software projects have been introduced, which extend the rapid application development methods developed earlier. The collective name for a group of those methodologies is Agile. The core of this method is the development of the client deliverable module and obtaining client acceptance. If necessary, the client may request additional changes before accepting the module. The same goes for the next module, etc.

The business process proposed in this book covers Agile application development as well. The module design is often done at a collective brainstorming session without having documented business requirements. After the module implementation and the first review with the client, a change request may be generated for required modifications. The process flow goes back from the Construction Frame to the Requirements Frame for establishment of the modified module requirements. When the module is completed and accepted, the process flow loops back from the Closing Frame to the Requirements Frame again to get a new set of requirements for another module. When using Agile development, there is much less emphasis on requirements management and planning and the project documentation is informal. The focus of the team is delivery rather than compliance with formal processes, and this methodology can indeed work in certain applications. While technically the project goals are usually achievable, the role of the project manager is reduced, the risk is higher and it is difficult to accurately predict cost and schedule for the entire project. Also, due to scarce documentation, the "lessons learned" analysis at the end of each module is not detailed enough and mistakes made in the project may be repeated all over again.

Workflow Overview

NOTE At this point it will be beneficial for the reader to revisit the section "Why This Book Is Different from Other Project Management Books" in the Introduction, where the concept of the PM Workflow framework is described, which is needed to understand the rest of the book.

The following is an overview description of PM Workflow, simplified to allow high-level understanding but which leaves out details to be covered later. It is very important to note that this is indeed an overview, meant to provide orientation and context. Later, the reader will be presented with greater details of each process flow and the sequence of processes in the project life cycle. The high-level overview project flow diagram is shown in Figure 1.1.

NOTE When reading process flow descriptions, it is advisable to have the process flow diagrams in front of you, via a printout or an easily accessible screen.

The workflow starts when an initial project request is received from a client in the Requirements Frame of the project. First, a Cost-Benefit Analysis (1) is performed to determine whether the project benefits justify the cost of the project. The benefits statement is usually provided by the client, while the cost-benefit analysis is often done by the project manager.

It is unlikely that a project's client will provide a detailed and complete set of project requirements. Therefore, the process flow is directed to the high-level process named Requirements Analysis (2) for the establishment of project requirements and analysis. This is a set of the major separately funded and planned detailed process activities. In fact, it has all the components of a complete project. The project requirements analysis, presented in the following chapters, describes methods of gathering and documenting business requirements from clients and the process flow to accomplish it. The process ends when a business requirements document (BRD) is signed by both the delivery organization and the client. This also implies that the budget, at least for the Planning Frame, is approved, since no project activities may take place without the approved budget.

Just after the requirements analysis is started in a new project, the traceability matrix does not yet exist, and the process flow is directed to the Create Traceability Matrix (3) process. The traceability matrix is used to document, update and trace all changes to requirements throughout the course of the project. The process of creation and maintenance of the traceability matrix defines methods of documenting requirements traceability. Think of the requirements traceability matrix as a way to be able to put a source to every requirement, answering the question "Who asked for this feature and why?"

The output of the new traceability matrix serves as the input for creation of the BRD in the Create Business Requirements Document (4) process. The BRD reflects only what has been agreed to in terms of the project functionality between the delivery organization and client before the detailed project planning takes place. The BRD is essential for presentation of the original scope to the project sponsor and getting budget approval for the agreed features of the project. The BRD contains only baseline requirements and will not be changed in the course of the project. All approved scope changes, even the minor ones, will be recorded in the traceability matrix, even when there is no impact on the project cost and schedule. Also, the project plan and schedule will be updated to schedule the implementation of those changes.

One of the elements of the BRD is the preliminary estimated budget required for the project. Once the BRD and the preliminary budget are approved, the Planning Frame begins. If the BRD or the budget is rejected, the project flow goes to the Closing Frame and the project is terminated. It is important to capture the project information even

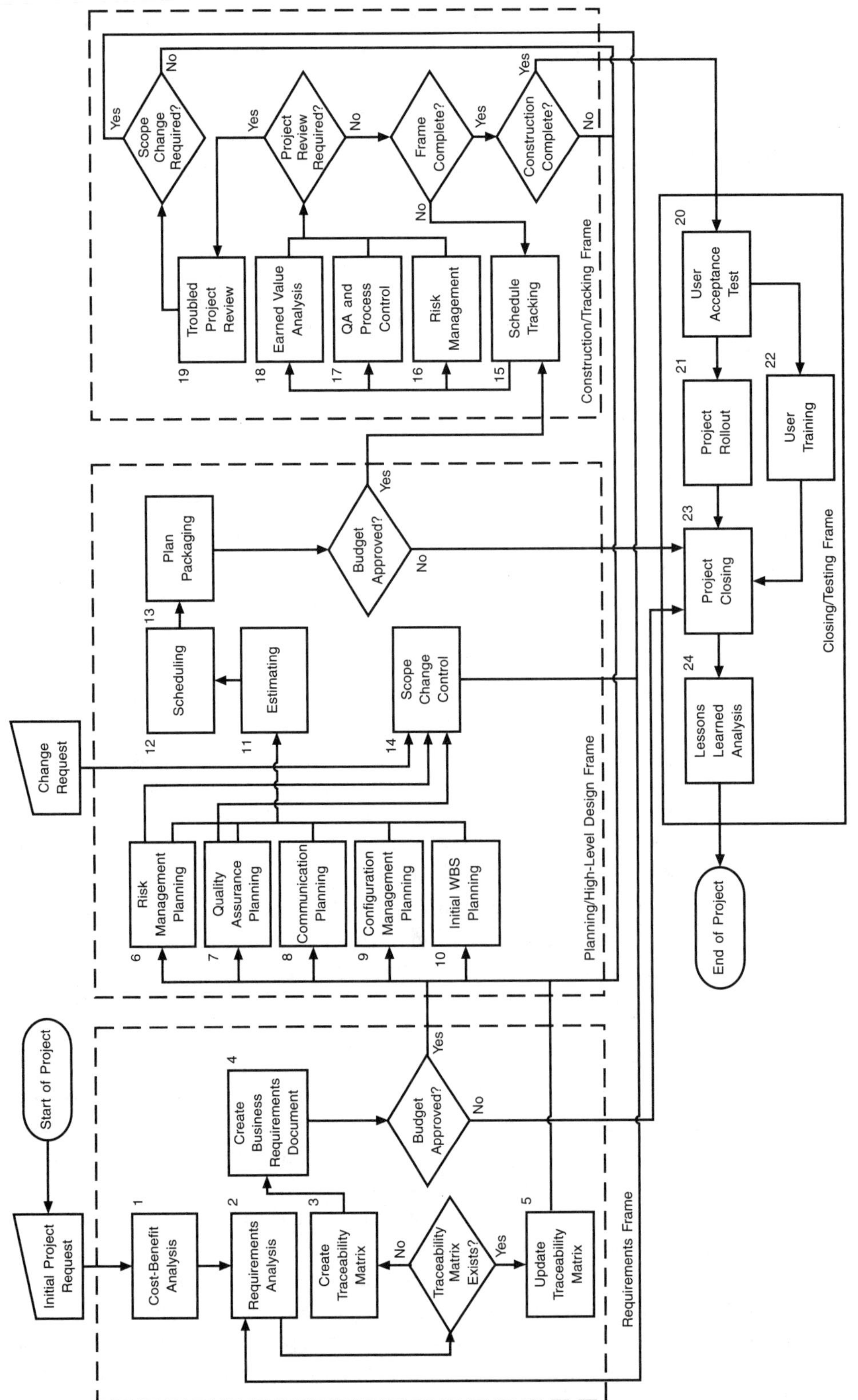

Figure 1.1 High-Level Overview Project Flow Diagram

in this case, for project lessons learned to assist in future project selections. As follows from the detailed review of the Requirements Frame process flow, the process allows for reducing the scope of the project before rejecting it. This may be useful if the issue is money. The client may still want the project delivered, but at the mutually agreed reduced scope.

Later, during the Planning or Construction/Tracking Frame, when a scope change request is produced, the process flow goes back to the Requirements Frame to analyze the scope change. When analysis of the change request is completed, the traceability matrix already exists, but it has to be updated in the Update Traceability Matrix (5) process due to the new scope change. The process flow is now directed to the Planning Frame for planning of the scope change implementation.

The purpose of the Planning Frame is to develop the complete detailed frame and project plan package, which includes the detailed project schedule, communication plan, risk management plan, quality assurance plan, configuration management plan and others.

Before development of the detailed project schedule, the following processes must be executed:

1. Risk Management Planning (6)
2. Quality Assurance Planning (7)
3. Communication Planning (8)
4. Configuration Management Planning (9)
5. Initial Work Breakdown Structure (WBS) Planning (10)
6. Other planning elements discussed in Section III on the Planning Frame

The completed detailed WBS and the plan provide the breakdown of all project tasks, their estimates, the task interdependencies and resources assigned to each task. In terms of the WBS process itself, it is a decomposition process so that larger, very complex deliverables can be described in smaller, more easily understandable and executable chunks (see the note about the WBS below). At the beginning of the Planning Frame it is not known yet how many resources are required and how many will be available for the project. The resource requirements will be determined after completion of the project estimates in the Estimating (11) process. As soon as the resource requirements are known, the resource requests will be submitted to the resource owners, and based upon available resources, the WBS will be updated in the Scheduling (12) process.

Formally speaking, the WBS is the decomposition of project activities, in which the project activities are listed in a structured, hierarchical, outline form to identify (at its top level) key work streams under the overall project deliverable, down to (at its lower levels) work packages which can be assigned to specific accounts and resources. However, *our* focus is process oriented. We add value by illustrating project workflow. The WBS cannot be called a process from the process modeling point of view, because a WBS by itself does not have a business *use* in and of itself. You cannot *control* a project with a WBS, because a WBS is only a *step* (*albeit a major one*) in creating a *process* which *does* have a business meaning.

> We therefore see the WBS as part of a process that includes the WBS corresponding estimates, task dependencies and resource assignments. This is described in detail in Section III on the Planning Frame. The resource assignment at this stage may not have actual resources assigned; generic names such as designer 1, designer 2, developer, programmer 1 and so on are used. This process is implemented when building a complete Gantt chart, which is a tool to plan the project and later to control it. At this point the process has a real business meaning, since it provides the visibility to the project implementation and may be used to obtain business consent. The resulting process is not a WBS by formal definition, nor is it a project plan. In order to create a project plan, many more project documents must be added, such as the communication plan, quality management plan, risk management plan and others. Some of the information from those plans may be entered into the Gantt chart, but much more is not. Thus, this process is *more* than a WBS, but something *less* than a project plan.
>
> In Section III, we call this process "WBS Design and Preliminary Project Planning (P5)." Since this process is mentioned in the book many times, it would be too cumbersome to refer to it by that lengthy name each and every time, so we took the liberty of calling this expanded WBS planning process simply the "WBS." Each time you encounter that name, remember that it is a more holistic and process-oriented view than the PMI and other traditional definitions of the WBS.

All plans and the detailed WBS will be put together into one project package in the process Plan Packaging (13) and sent out for budget approval by the project sponsor. If the budget is not approved, the project will be terminated. Otherwise the schedule tracking starts in the Construction Frame in the process Schedule Tracking (15) and from there as described below in this section.

If a project scope change is initiated at any point in the project life cycle, after initial assessment in the process Scope Change Control (14), the project flow is directed back to the Requirements Frame to do the scope change Requirements Analysis (2) and then to the Update Traceability Matrix (5) process in order to update it. From there the flow is directed to the Planning Frame again, which again invokes some or all of the processes (6 through 10) at the start of the Planning Frame, depending on specific circumstances. Since most scope changes involve additional work, the extra budget and new completion date must be approved before the implementation. This is critical. It is the omission of this step that leads to many project failures. Scope changes must be formally accepted and their implications to the project's schedule and budget must be fully and thoughtfully considered. A scope change will be implemented in the Construction Frame. In fact, the Scope Change Control (14) process provides more functionality and controls more processes than shown in the simplified process flow diagram. Details will be discussed in Section III on the Planning Frame.

During the Construction/Tracking Frame the schedule is tracked in the process Schedule Tracking (15), where the actual cost and schedule are compared to what was planned. The Schedule Tracking process utilizes processes 16, 17 and 18. In order to control project financials, the process Earned Value Analysis (18) is used. The Quality Assurance (QA) and Process Control (17) process ensures that all project management processes are being followed. Periodic risk assessments are performed in the process Risk Management (16). If serious issues are discovered in the project implementation, such as late schedules, overspending, quality issues, new serious risks, etc., the process QA and Process Control (17) will direct the process flow to the process Troubled Project Review (19) to determine the reason for unsatisfactory performance and develop a plan to remediate it. If it is determined that in order to resolve the problem the project scope has to change, the process flow will go back to the Requirements Frame to come up with the scope change requirements. Otherwise, it will loop back to the Planning Frame for adjustment of the existing plan to remediate the current situation.

Why is it necessary to go back to the Requirements Frame for a scope change analysis if there are no issues with the scope, but the project is late or over budget? The reason is based on the concept of the triple constraint. This is the ongoing balance between scope, time and cost. Although this specific terminology is no longer explicitly mentioned in the *PMBOK® Guide*, it is still a very valuable lens with which to view a project. The constraints are interdependent, so if the project is late and/or over budget, there will be pressure to reduce scope. The client will insist on delivery dates as scheduled and no additional budget will be available. In fact, *the client's* business may suffer monetary losses if the project is not completed on time. Clients would rather agree on temporarily reducing the scope of the project with some elements still missing than having nothing delivered on the due date. They may agree that missing project elements will be delivered in a new yet undocumented future phase 2 of the project. Both the client and the project manager hope the budget issue will be resolved by then, even though in reality this sudden phase 2 rarely occurs. If the project satisfies basic business needs without some features, no budget will be available for phase 2. Rather, if several additional, previously unplanned modifications are required for the project after it is delivered to the client, a new project will be created for them. Clients will tend to include in this new project the missing features from the completed project with the reduced scope.

It will be shown later in detailed process flow descriptions that every frame implementation is tracked in the Construction Frame, including the construction part of the Construction Frame. If the tracking of any frame processes is complete, the "frame complete" test condition will be true, which will direct the project flow back to the Planning Frame for further planning or modification of the project activities. After completion of each frame, except the Closing Frame, the WBS must be modified to reflect new available project details. The project plan for the next frame must be updated, the risk analysis performed, new estimates made, the next frame budget approved and so on.

If the "construction complete" test is true, the entire project construction is complete and the process flow is directed to the Closing Frame. At that time the User Acceptance Test (20) is conducted and the project is rolled out to the client in the process Project Rollout (21). Then the Project Closing (23) is executed and all resources are released. At the same time as the user acceptance test starts running, User Training (22) is provided.

The last process for all projects is the Lessons Learned Analysis (24), which analyzes what went well and what errors were made in the project, with recommendations for how to avoid similar errors in the future. Although the workflow shows this as the "last process," it is important that the team record lessons documented and learned as they happen—when the pain (or pleasure) of the events are fresh and the information is fresh. You will see requests to document all project events throughout this book.

NOTE The overview process flow diagram (Figure 1.1) does not show the complicated nature of interconnection of the frames. In order to see all interconnections, refer to the frames interaction diagram in Appendix B.

Project Sequence Diagram

The overview project management process flow diagram just scratches the surface of the amount of interaction between project frames. Each project frame has three elements, which are not necessarily performed in that frame:

1. Frame planning
2. Frame implementation
3. Frame tracking

Every project must have a plan, which describes what will be done—and how, when and by whom—in the project. Once the plan is available, the plan implementation starts, at the same time tracking it and ensuring that the project is on schedule and within the budget.

When looking back at the overview process flow, you may see that the purpose of the Requirements Frame is to gather and document business requirements in the BRD. It does not have processes for planning or tracking the implementation. However, the Planning/High-Level Design Frame has planning processes. The Construction Frame has processes required for tracking of the frame implementation. Therefore, soon after starting the Requirements Frame, the project flow is directed to the Planning/High-Level Design Frame to develop the requirements plan. After developing the plan, the flow goes back to the Requirements Frame to start the plan implementation. Concurrent with the beginning of the plan implementation, the process flow goes to the Construction Frame to start tracking the implementation of the Requirements Frame.

It is not always possible to track all processes, because tracking involves comparison of actuals with the plan or the established completion criteria. For example, when planning initial processes of the Requirements Frame, there is no plan yet and therefore it cannot be compared to actuals. However, cost estimates for those initial processes must be available and adhered to. By the time the requirements analysis planning is complete, it will be tracked in the Construction/Tracking Frame.

Table 1.1 describes the three elements of each frame and points out the frame where each element is executed.

The overview project frame sequence diagram describes the process flow between project frames in a mode when there are no issues with the project and no scope changes are requested. While this is an unlikely scenario, the purpose of this sequence diagram

Table 1.1 Frame Activities by Location of Execution

What Is Done	Description	Where It Is Done
Requirements Frame planning	◆ Plan gathering and documenting business requirements	Planning Frame
Requirements Frame implementation	◆ Generate requirements ◆ Develop BRD ◆ Develop traceability matrix	Requirements Frame
Requirements Frame tracking	Track implementation of the: ◆ Requirements generation ◆ BRD development ◆ Traceability matrix development	Construction Frame
Planning Frame planning	◆ Plan the high-level design ◆ Development of the schedule and project plan package	Planning Frame
Planning Frame implementation	◆ Perform high-level design ◆ Develop intermediate tasks plan, such as WBS, estimates, risk analysis, quality, etc. ◆ Develop project plan package	Planning Frame
Planning Frame tracking	◆ Track high-level design implementation ◆ Track intermediate tasks ◆ Track project plan package development	Construction Frame
Construction Frame planning	◆ Plan detailed design and implementation	Planning Frame
Construction Frame implementation	◆ Perform detailed design and implementation	Construction Frame
Construction Frame tracking	◆ Track detailed design and implementation	Construction Frame
Closing Frame planning	◆ Plan user acceptance test ◆ Plan project closing	Planning Frame
Closing Frame implementation	◆ Implement user acceptance test ◆ Close the project	Closing Frame
Closing Frame tracking	◆ Track user acceptance test ◆ Track project closing	Construction Frame

is to illustrate the high-level frame flow and to provide context. Later, other situations will be reviewed, including when a scope change request is issued because the project is late or over budget or because other major issues are discovered.

After receiving an initial project request in the Requirements Frame, the process flows to the Planning Frame in order to convert these requirements into an actionable plan. The major planning activity for the Requirements Frame is the planning of the project requirements analysis. All activities during the project life cycle first must be planned and only then executed according to the plan, along with the tracking of the execution by comparing the actual performance elements of the project (such as schedule, cost and quality) against the planned ones.

Planning of all frames is always done in the Planning Frame, because it is equipped with the tools, methods and processes necessary for planning, estimating and approval of the plan. Tracking is always done in the Construction Frame for the same reason.

Upon completion of the Requirements Frame planning, the process flow returns to the Requirements Frame to implement the Requirements Analysis process, at the same time monitoring and tracking it in the Construction/Tracking Frame. Both the Requirements Analysis and the Requirements Analysis Tracking processes start at the same time, run at the same time and end simultaneously. When both of these processes are about to approach their final days, a new process starts in the Planning Frame for planning of the high-level design and planning of the Planning Frame tasks, which runs simultaneously with the requirements analysis. The Requirements Frame ends when the project sponsor authorizes further project development beyond the requirements analysis. The frame sequence flow diagram is shown in Figure 1.2.

Planning is required for the Requirements, Construction and Closing frames. However, clarification is required in order to understand the concept of "planning for the Planning Frame." If the frame had the name High-Level Design Frame, there would not be that much fuss about planning it, but planning of the Planning Frame may sound redundant or confusing.

Let's look in detail at the Planning Frame using the example of a software project. Here, the Planning Frame includes development of the high-level system architecture and the overall design, which will have established a list of software modules in the application being developed, program architecture, specific inputs, output and processing, as well as interaction with other software modules and even different applications. The second part of the Planning Frame is the plan package, which includes estimates, risk analysis, quality tasks, cost, duration, etc. The overall plan is a detailed plan for the Planning Frame and a high-level project plan to the end of the project, which includes the Construction Frame and Closing Frame plans. While the overall plan is not intended to be very detailed at this point, it provides an idea of and a starting basis for the overall project cost and duration. This plan is not intended for execution, because it is not detailed enough. This plan is required to receive authorization from the project sponsor to continue the project beyond requirements analysis. The plan for the Planning Frame may include detailed plans for tasks like "Run Risk Assessment" or "Perform Quality Assurance Review," while the implementation of those tasks would actually imply running risk assessment and performing quality reviews.

As soon as the requirements analysis is done and the high-level design plan is complete, the implementation and tracking of the high-level design starts. Later, during the high-level design implementation and tracking, the development of the Construction Frame plan starts. By the end of the high-level design implementation, the Construction Frame plan will be completed.

Similarly, as soon as the Planning/High-Level Design Frame is complete, the implementation and tracking of the Construction Frame starts. Before the end of the Construction Frame planning, development of the Closing Frame plan starts. By the end of the Construction Frame implementation, the Closing Frame plan will be completed.

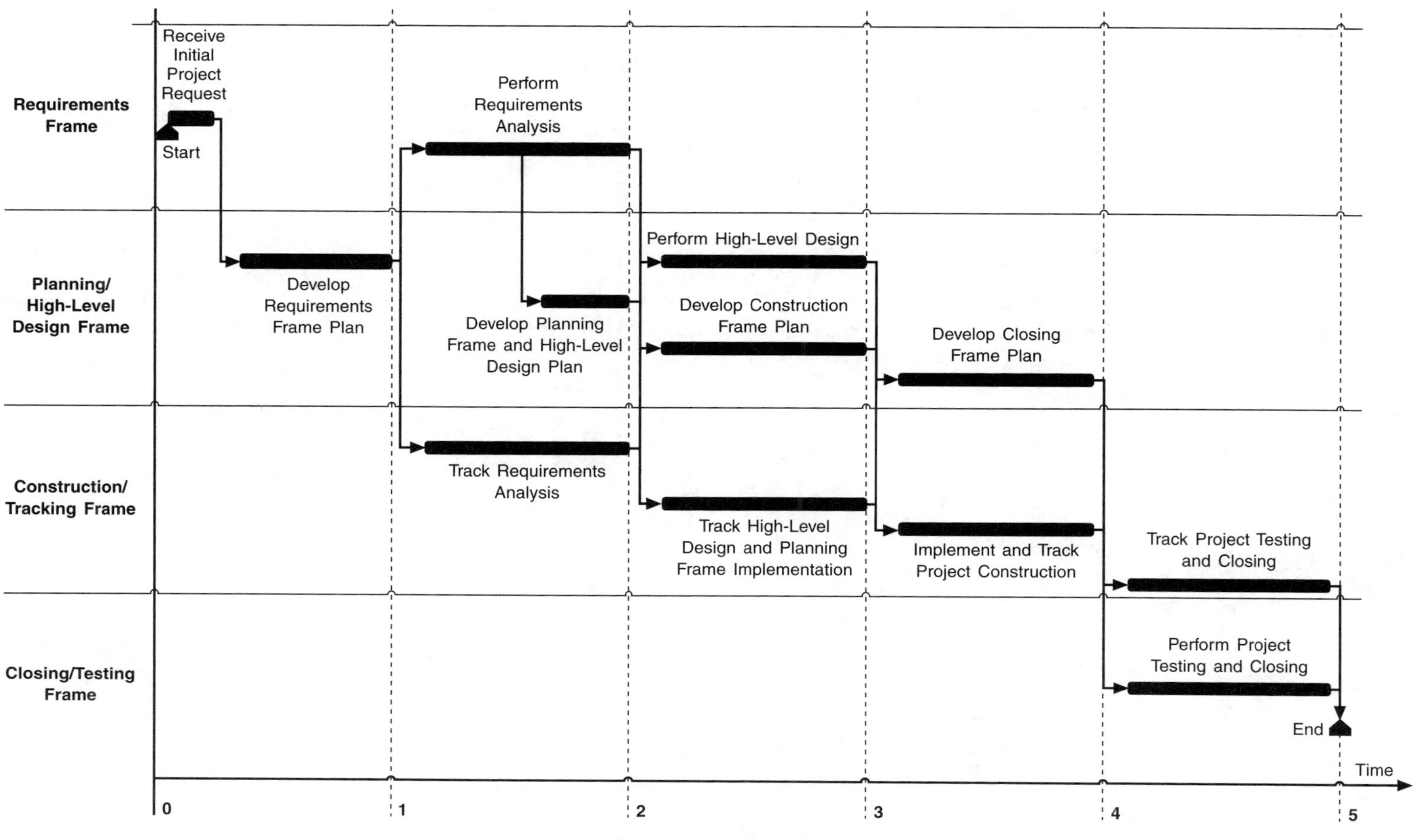

Figure 1.2 Frame Sequence Diagram for the Ideal Project

When the Construction Frame is complete, the Closing/Testing Frame plan is also complete, and the project starts user acceptance tests and closing out. Simultaneously, the tracking of the testing and closing is performed.

Process Naming Convention

All the project processes described in this book have a unique identifier. Thus, for requirements management processes in the Requirements Frame, the unique identifier starts with letter R (for **R**equirements), followed by one or more numbers. If the process is a top-level process in the Requirements Frame, the letter R is followed by a number, as in R4, R5, etc. If process R4 is further decomposed, additional numbers are appended after a hyphen, like R4-2. Further breakdown is identified with more hyphens and numbers, for example R4-2-1.

All detailed process names start with one of following characters, according to the project frame:

R Requirements management processes in the Requirements Frame
P Planning processes in the Planning/High-Level Design Frame
C Tracking processes in the Construction/Tracking Frame
T User acceptance test and closeout processes in the Closing/Testing Frame

All processes as marked in the process flow diagrams are indicated by a number in a circle (see the example in Figure 1.3).

In several instances, processes are denoted with additional characters, either "a" or "b," as in R5a and R5b. This is the same process which is executed in the same frame but for different purposes. For example, process R5 is used for creation of the traceability matrix (R5a) and for its update (R5b). Usually it is executed once for the former and many times for the latter. Because the conditions under which process R5 is executed are different, processes R5a and R5b are located in different areas of the process flow diagram.

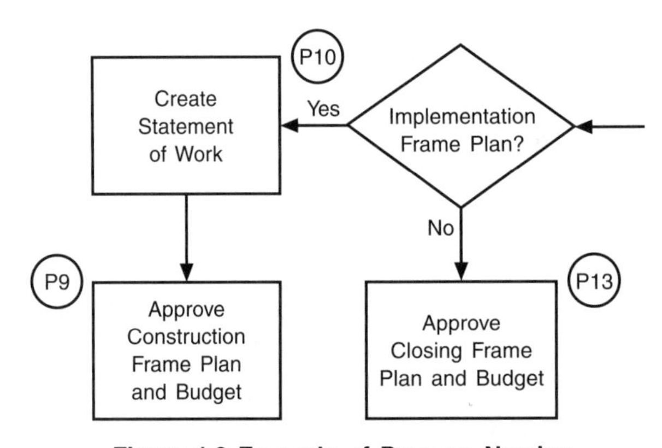

Figure 1.3 Example of Process Naming

All entry and exit points to and from each frame have a number, sometimes followed by a letter. In order to show the connection between frames, the same exit point number identifies the corresponding entry point in another frame. For example, exit point 1 in the Requirements Frame is connected to entry point 1 in the Planning Frame. The letter that follows the number identifies the specific reason for the exit. Exit points 1A, 1B, 1C, etc. are connected to the same entry point in the Planning Frame. However, 1A means the project flow goes from the Requirements Frame exit point 1 to the Planning Frame entry point 1 for Requirements Analysis Planning (1B) for the scope change plan and so on. The letters have no further meaning and are provided for clarity only.

Section II.
Requirements Frame

Introduction and Initial Project Analysis

Overview of the Requirements Frame

The main objective of the Requirements Frame is to ensure that the project scope and/or scope change requirements are agreed on and documented and the requirements baseline is produced.

The term *project scope* refers to the scope of work required to deliver the solution to the client in accordance with the *product* scope. Product scope, in turn, is the scope of the business solution, based on business requirements. Any change in business requirements will lead to a project scope change. Therefore, the term *project scope change* is also reflective of business requirements changes.

Requirements management controls the flow of customer requirements through the life cycle of the project to ensure understanding of and agreement with the scope of the project by the delivery team and business. Also, it provides traceability of changes to the project scope to ensure that the project delivers exactly the product scope agreed upon with the client.

The requirements management process consists of the following major elements:

- Receive initial project requirements and a project benefit statement from a client.
- Perform cost-benefit analysis and justify the project investment.
- Create project control book to establish guidelines and tools for documenting requirements and project events.
- Identify project stakeholders and establish communication channels with them.
- Elicit detailed business requirements for the project.
- Conduct business requirements analysis.

- Create business requirements document.
- Analyze project scope change requests.
- Create/update traceability matrix for requirements, which allows monitoring requirements changes throughout the project life cycle.
- Obtain authorization to start the project.
- Obtain project funding.

The requirements management set of processes also includes gathering and documenting business requirements and reviewing them for completeness, clarity, understanding, prioritization, testability and approval. Changes to the product scope or requirements are only executed through a project scope change request, described in Chapter 10. All changed requirements are documented in the traceability matrix as revisions or updates to the baseline, as well as in the project control book.

Roles and Responsibilities

The following roles will be required to execute the Requirements Frame:

- **Project manager**—Responsible for ensuring that all processes of the workflow are adhered to and documented in the project control book. The project manager's signature is required on the business requirements document. The project manager tracks the implementation of the requirement management activities and reviews the status of the Requirements Frame execution on a weekly basis with the lead client.
- **Delivery team**—The group of project team members. They participate in the requirements review sessions to make sure that each requirement is clear and technically feasible to implement and test. They will be responsible for producing documentation in the Planning Frame and outlining how the approved requirements will be implemented.
- **Requirements manager**—Responsible for documenting and managing the requirements throughout the Requirements Frame of the project. This person may be a project manager or a senior business analyst. The requirements manager's signature is required on the business requirements document.
- **Lead client**—The client responsible for presenting business needs to the project manager, reviewing requirements documentation and assuring that senior business managers and the project sponsor are in full agreement with the proposed requirements. The lead client is assigned to work with the requirements manager as planned by the project manager and participates in all relevant business requirements gathering sessions. The lead client sometimes is referred to as the business project manager or the business area lead. The lead client usually reports to the senior business manager.
- **Business expert**—The business user from one or more business areas on whose behalf the specific business requirements are presented. The business expert is a

consultant to the lead client and to the project manager/requirements manager for that business area.

- **Senior business manager**—The person whose signature is required on the business requirements document. The senior business manager ensures that business requirements are aligned with the corporate business goals and corporate strategy. The senior business manager usually reports to the project sponsor.
- **Senior delivery manager**—The person who owns the delivery budget and whose signature is required on the business requirements document. The project manager usually reports to the senior delivery manager, at least for the duration of the project.
- **Project sponsor**—Major stakeholder who is responsible for the business success of the project, specifically, ensuring that the business objectives for which the project has been undertaken are met. The project sponsor is the owner of the overall project.
- **Quality assurance analyst**—Responsible for running requirements reviews, documenting results of reviews and monitoring follow-ups when needed. Ideally, the quality assurance analyst should be a person with quality assurance training who is not directly associated with the project. However, for smaller projects, the project manager may play this role.

NOTE This nomenclature is provided here to align the reader for our treatment of these roles in the project workflow. Your organization may have different titles but should have very similar roles to which you can draw analogies and more clearly follow the workflow.

Inputs/Outputs

The project process flow enters the Requirements Frame and exits it depending on the specific needs of the project through the following entry points (see the Requirements Frame process flow diagram in Figure 2.1):

- **Entry point 4**—From the Planning Frame after completion of the ballpark or initial estimates (estimates made after the initial project request)
- **Entry point 5**—From the Planning Frame to request requirements analysis for a scope change
- **Entry point 6**—From the Planning Frame after the requirements analysis plan is finished
- **Entry point 8**—From the Planning Frame with the Planning/High-Level Design Frame plan
- **Entry point 9**—From the Planning Frame with preliminary estimates (first estimates made after the business requirements document is complete)

The process flow exits the Requirements Frame through the following exit points:

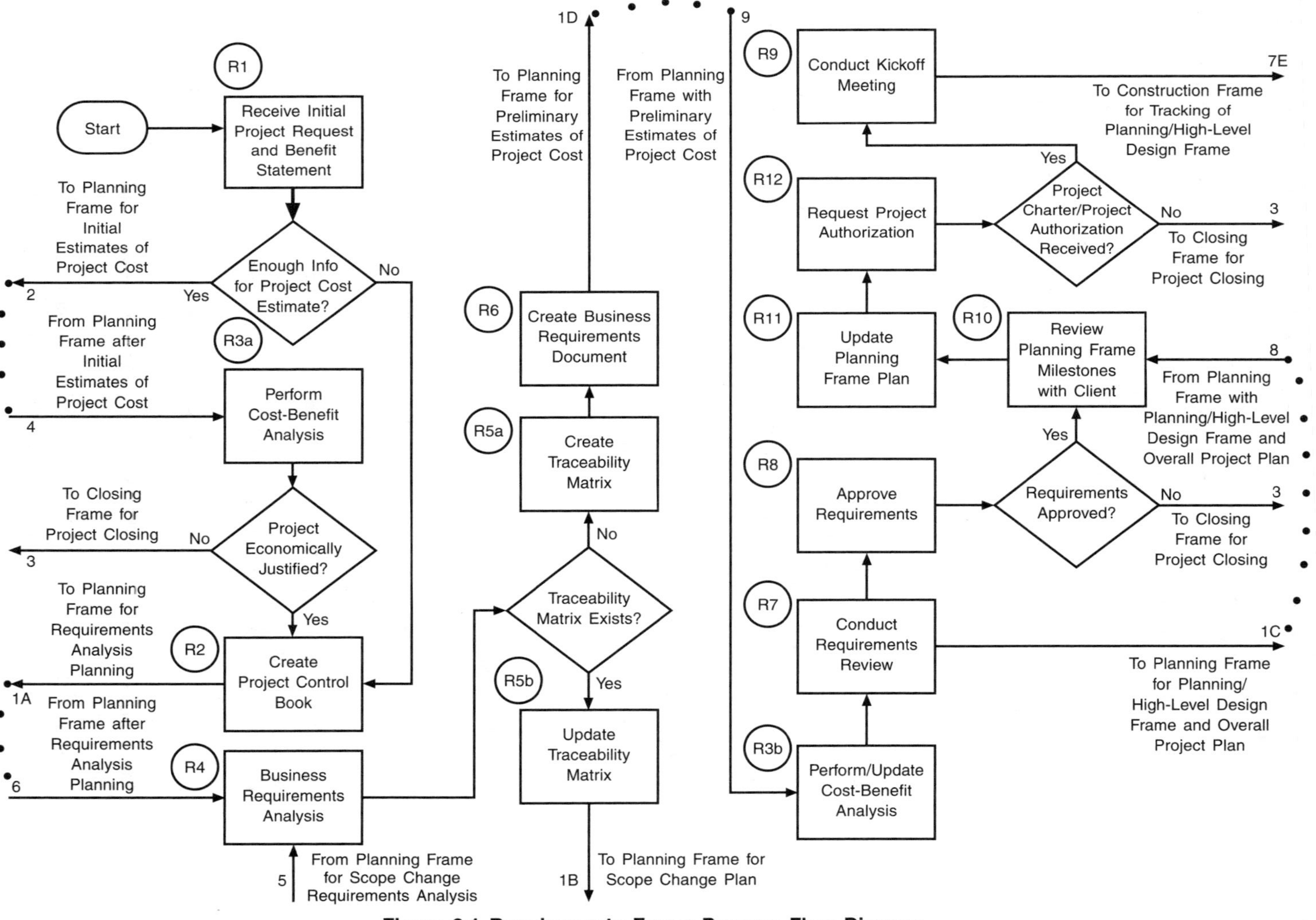

Figure 2.1 Requirements Frame Process Flow Diagram

- **Exit point** 1—
  - ◇ 1A—To the Planning Frame requesting the requirements analysis plan
  - ◇ 1B—To the Planning Frame requesting the scope change plan
  - ◇ 1C—To the Planning Frame requesting the planning/high-level design and the overall project plan
  - ◇ 1D—To the Planning Frame requesting the preliminary estimates of the project cost
- **Exit point 2**—To the Planning Frame requesting the ballpark estimates
- **Exit point 3**—To the Closing Frame requesting project termination
- **Exit point 7E**—To the Construction Frame requesting the Planning/High-Level Design Frame implementation and tracking

Requirements Frame Process Flow

The Requirements Frame process flow consists of the following processes:

1. **Receive Initial Project Request and Benefit Statement (R1)**—This process will provide the delivery team with a general idea about the project and will outline expected benefits from the project.
2. **Create Project Control Book (R2)**—This process will create a tool for keeping all project documentation in one place.
3. **Perform/Update Cost-Benefit Analysis (R3a and R3b)**—Cost-benefit analysis calculates whether benefits from the project justify expenses.
4. **Business Requirements Analysis (R4)**—This process will elicit detailed project requirements and analyze them.
5. **Create/Update Traceability Matrix (R5a and R5b)**—The requirements traceability matrix is a tool for documenting, updating and tracking all requirements and changes to the project scope throughout the life of the project.
6. **Create Business Requirements Document (R6)**—This process will create a document outlining the baseline for all project requirements.
7. **Conduct Requirements Review (R7)**—This process will ensure the quality of business requirements and verify correctness of the business requirements document.
8. **Approve Requirements (R8)**—This process will approve requirements and authorize funds for the Planning Frame.
9. **Conduct Kickoff Meeting (R9)**—The kickoff meeting is the official project initiation.
10. **Review Planning Frame Milestones with Client (R10)**—This process will review the planning/high-level design and the overall project plan or milestones with the client before requesting project authorization. Plans may be changed later as a result of the review.
11. **Update Planning Frame Plan (R11)**—This process will update plans or milestones with changes requested at the review before beginning the Planning/High-Level Design Frame implementation in the Construction/Tracking Frame.

12. **Request Project Authorization (R12)**—This process will request project approval and the project charter from the project sponsor. Receiving a signed project charter confirms that the sponsor agrees with the project scope, including the cost and schedule of at least the planning/high-level design activities for the remaining frames of the project.

After receiving the initial project request and benefit statement (R1), the next task of the Requirements Frame is to provide the ballpark estimates of the project cost, assuming that sufficient project information is available from past experience to do this sort of estimation.

Process R1 needs to establish whether it is possible to perform the ballpark cost estimates without having detailed project requirements beyond the initial project request. This may not be possible unless (or until) the delivery team has significant experience in the same type of projects. Since costs of similar past projects are known, it is possible to establish the approximate cost of new projects which are similar in project context. If the project is in an unfamiliar business or a different or brand new technology, then the detailed project requirements must be established first. It is also critical to know whether or not the *delivery* team has significant experience in delivering similar projects. Even if they do, the cost may only be roughly estimated, consistent with other completed similar projects, making corrections as appropriate. The lowest estimating accuracy allowed for the ballpark project estimates is –25% to +75%. If this accuracy is not achievable, no estimates will be made at this point and the next process in the process flow, Create Project Control Book (R2), will be executed.

If there is enough information for the ballpark project cost estimates, the process flow is directed to the Planning Frame through exit point 2. The estimating process is located there since the estimating will be done in the Planning Frame many times during the project life cycle. This process is covered in Chapter 11. After receiving the ballpark cost estimates from the Planning Frame via entry point 4, the process Perform Cost-Benefit Analysis (R3a) is initiated. If it is not possible to estimate the project at this time, skip the estimating and the cost-benefit analysis for now and do them later when it is possible.

In the case where the project is the result of a winning bid, then instead of the initial project request, there will have already been a winning proposal submitted by one of the bidders to the client. In this case, no cost-benefit analysis is required. The project was awarded to one of the developing organizations, because the client found its proposed solution and the price to be the most cost effective.

Proposal writing is out of scope for this book, since it relates to the subject of proposal management rather than project management. Even though proposal managers are usually experienced project managers, the spectrum of issues associated with proposals is not the same set of issues faced by project managers.

As mentioned earlier, the initial project request almost never provides detailed specifications that can be used to build detailed project plans. At this stage, users often are not even sure about what they want—at least not *exactly*. Sometimes mutually incompatible requirements come from several business user groups. In order to come up with the set of requirements, agreed on by all parties, including the delivery team, a costly analysis is often required here. Before committing to pay for this work, the client may want to know whether expenses justify expected revenues from the project implementation. The projected revenues are provided by the client in the benefit statement, which is a part of the initial project request. Thus, the client may estimate that the project implementation will provide revenue of, say, $2 million during the first year, after which it will increase to an annual $5 million each of the following three years. If the cost of the project is estimated at $7 million, then the money spent will be recouped in two years, without accounting for the effects of interest rates and inflation. It will be a business decision whether it makes sense to wait two years before getting any profit from the project. Details of cost-benefit analyses will be provided later in this chapter.

The general rule is if the revenue generated by a project does not cover the cost of the project in 2.5 years after project completion, a commercial project is not justified. The actual number will, of course, vary for different organizations and projects. As a business owner, the client should know how the business is expected to profit from the project and provide this information in the benefits statement. Based on this and the ballpark project estimates, the information needed to perform the cost-benefit analysis is available. If project benefits do not justify project spending, the project should be terminated and the process flow through exit 3 is directed to the Closing Frame for project termination.

In some cases, however, the project is justified *even if no profit is expected.* Let us say, for example, that you work for a bank, which has just received information about new government taxation rules effective three months from now. In this case, the bank must start a project to modify software programs that calculate taxes by the due date, even though there will not be any financial benefits to the bank from doing that. The option of ignoring this requirement may be too costly for the bank or may prevent it from staying in a profitable line of business. This is an example of a project undertaken for economic sustainability.

Another example is when a highly visible project is not expected to be financially beneficial, but success would improve the future business situation. Therefore, project expenses may be regarded as an investment. Again, this is an example of investing for the longer term—for economic sustainability.

If it is determined that the project is justified, it is necessary to establish detailed business requirements as accurately as possible. If it is not clear what precisely should be delivered to the client, then it is not possible to know how long it is going to take and how expensive it is going to be, let alone the issue of client satisfaction.

NOTE Often you will be under severe pressure from your management to provide project estimates, because your client is reluctant to commit funds even for the requirements analysis if they have no idea about the total project cost. If you submit to the pressure and provide estimates for the entire project, which may later turn out to be twice as high, there will be severe consequences for you, because your management's insistence on providing project estimates without sufficient information will not be remembered. The opposite situation, when your estimates are way too high, won't be any better for you. Usually, the company budgeting process involves the most senior managers, who allocate budgets for all projects based on the available funds. Your failure to provide reliable estimates may cause them to change the budgeting of all projects in the organization. For sure, someone will request that this never happens in the future, which will not end well for you!

If the cost-benefit analysis confirms that the project is economically justified, a tool must be created in process R2 for the storage of all project documentation. The tool is called a project control book. The golden rule for the project documentation is that if anything during the project life cycle is not documented, it is the same as if it does not exist or never happened. Phone conversations, verbal agreements and promises do not substitute for documentation, since management or clients will never remember their undocumented requests or their consent to do something. Methods of creating the project control book for documenting all project events are described later in this chapter.

After the process Create Project Control Book (R2) is completed, the process flow goes to the Planning Frame for the Requirements Frame planning using exit point 1A. The planning process is covered in Section III on the Planning Frame.

Just as with all project activities, the Requirements Frame must be planned in the Planning Frame of the project. Planning ensures that the Requirements Frame of the project proceeds according to the schedule and has a predictable cost. The Requirements Frame tracking (plan versus actuals), status check and reporting are done in the Construction Frame. Although many elements of the generic plan will be discussed in the Planning Frame, some specifics for the Requirements Frame planning are included here.

The Requirements Frame plan includes the following:

1. Analysis of all requirements
2. Identification of all deliverables
3. Documentation of the Requirements Frame team structure
4. Documentation of assumptions, dependencies and constraints
5. Analysis and documentation of the user environment factors, like physical work environment, technology used, users' levels of computer literacy, level of business expertise and training needs and their impact on requirements
6. Development of the risk management plan in accordance with the risk management procedure described in Chapter 6; the risk management plan will include, along with the planning of standard risk assessments and risk handling, the following:
    a. Development of a plan to minimize the risk of management or client pressure to limit analysis and start development as soon as possible
    b. Development of a plan to minimize the risk of neglecting nonfunctional requirements, like usability, training, etc. (this is relevant mostly in projects for developing machinery and software)
    c. Development of a plan to minimize the risk of technical specialists' bias toward a specific product, service or process

7. Development of a communication plan, which identifies stakeholders, reporting, distribution, etc.
8. Development of a detailed list of tasks (work breakdown structure) and dependencies between tasks
9. Estimation of effort required to complete each task
10. Identification of the resources required to complete Requirements Frame tasks and the requirements-related Planning Frame tasks and assigning resources to every task
11. Development of a quality assurance plan
12. Calculation of the cost of and producing the schedule for the Requirements Frame
13. All plans and the schedule combined into one Requirements Frame plan package
14. Getting to know clients, business users and project sponsors

The requirements manager has responsibility for planning activities. The project manager should include this plan as a component of the overall project plan. The Requirements Frame plan must make visible all activities and scope of the planned work, which would allow for correctly predicting time and cost of the requirements management activities.

When the process flow returns back to entry point 6 from the Planning Frame for the very first time in the project life cycle, this indicates that planning of the Requirements Frame has been completed and the Business Requirements Analysis (R4) process can start. During the course of the project, the flow may also be directed to the Business Requirements Analysis process through entry point 5 for planning of a project scope change. This is covered in Chapter 10.

The purpose of the Business Requirements Analysis process is to elicit detailed requirements from clients and business users, as well as to control the flow of customer requirements through the life cycle of the project to ensure understanding of and agreement to the scope of the project between the delivery team and client.

Methods to elicit detailed requirements from clients are described in Chapter 13. The requirements, as gathered, must be documented in the traceability matrix. Directly after the requirements analysis planning is complete, the process flow returns from the Planning Frame to the Business Requirements Analysis process via entry point 6. At this point the traceability matrix is not yet available. It is now time to build this in the Create Traceability Matrix (R5a) process. Based on the traceability matrix, the first client deliverable document can be produced in the process Create Business Requirements Document (R6), which provides details of all business requirements. While the business requirements document never changes, because it is a baseline set of requirements, the traceability matrix reflects the current status of requirements; therefore, it changes during the course of the project. It contains, along with the baseline requirements, the history of all changes to requirements during the project life cycle.

When a scope change is initiated at any time during the project life cycle, the process flow will go to the Business Requirements Analysis process through entry point 5. Since by now the traceability matrix already exists, the answer to the control point question (traceability matrix exists?) is yes. After completion of the scope change requirements analysis, the traceability matrix will be updated in the process Update Traceability Matrix (R5b) and the process flow goes through exit point 1B to the Scope Change Control (P7) process in the Planning Frame for planning the scope change and then executing it in the Construction Frame.

Having detailed requirements documented in process R6, the preliminary cost estimates will be performed, which are more accurate than the ballpark (initial) estimates. Why bother doing initial estimates earlier in the project if the preliminary estimates are now made with better accuracy? The answer is *cost savings*! When the initial estimates are done earlier and it is established that the project is not justified, there would be no need for expensive requirements analysis, planning and scheduling.

The process flow enters the Planning Frame for preliminary estimates of the project cost through exit point 1D. When it comes back with preliminary estimates through entry point 9, the next step is to do the cost-benefit analysis. This will be done in the process Perform/Update Cost-Benefit Analysis (R3b), provided it was not done earlier at step R3a or updated it if was. Having done so, the process flow enters the process Conduct Requirements Review (R7). Process R7 has many commonalities with the quality assurance review described in Chapter 14, but it is much simpler, with only one type of review instead of several types of quality reviews for other project frames. Therefore it makes sense to have a separate process R7.

During execution of the process Conduct Requirements Review (R7), when it is close to completion, the process flow is directed back to the Planning Frame again via exit point 1C, this time to produce a detailed plan for the Planning Frame, based on the gathered requirements. A less detailed plan for the rest of the project is also produced. When the planning is done after the business requirements document is complete but before requirements approval, the plan may change. However, it still provides a good picture of the Planning Frame cost and schedule. Corrections will be made later, if necessary, in the process Update Planning Frame Plan (R11).

Once the requirements review is complete, the new business requirements document, which also contains results of the cost-benefit analysis, will go for approval to the process Approve Requirements (R8). The approval will ensure that the analysis is complete and that the business requirements, which are supplied by the business, are presented correctly in the business requirements document, which is signed off on by both the delivery team and business management and is also documented in the project control book.

When the Planning Frame plan is complete, the process flow returns back from the Planning Frame to the Requirements Frame via entry point 8 in order to review plans, costs and schedule with the client in the process Review Planning Frame Milestones with Client (R10). Many plans are produced in the Planning Frame, which take significant effort, time and cost. The plan and resource allocation must be available for project estimating, the risk analysis, quality management planning and other elements of the Planning Frame. While the complete plan is mostly a guide to the delivery team, some elements, like major milestones and the cost estimates, will, of course, be presented to the client. As the result of the above-mentioned review, some changes may be introduced into the plan in the process Update Planning Frame Plan (R11). At this point it is assumed that the business agreed with the scope of the project, its cost and schedule, and the formal project approval request is forwarded to the project sponsor in the process Request Project Authorization (R12).

Based on the client's input and budget considerations, the project sponsor and/or client management has to produce the project charter, which is the authorization to start the project. If the project authorization is not provided by the client's management after several requests, the answer to the control point question (project charter/project authorization received?) is no, and the process flow is directed to the Closing Frame for project termination via exit point 3. Otherwise, if the answer is yes, the kickoff meeting (R9) will take place with the participation of the client, client management, the delivery team and delivery managers. After the kickoff meeting is complete, the process flow enters the Construction Frame for tracking of the Planning/High-Level Design Frame via exit point 7E. The tracking process is described in Chapter 15.

It was mentioned earlier that the process flow exits the Requirements Frame for several different reasons using exit points 1A, 1B, 1C, 1D and 2. All those points are connected to entry points 1 and 2 correspondingly in the Planning Frame process flow diagram (see Figure 5.1). Summarizing those reasons, the process flow is directed from the Requirements Frame to the Planning Frame for the following requests:

1. Ballpark estimates of the project cost (exit point 2)
2. Requirements analysis plan (exit point 1A)
3. Project scope change plan (exit point 1B)
4. Planning/high-level design plan (exit point 1C)
5. Preliminary estimates of the project cost (exit point 1D)

The process flow may also move from the Requirements Frame to the Closing Frame when it is determined that the project is not justified or when the project sponsor declines to issue the project authorization (in both cases via exit point 3).

Receive Initial Project Request and Benefit Statement (R1)

The project manager receives the initial project request and the benefit statement from the client. The resource manager documents them in sufficient detail to ensure their unambiguous understanding. A unique identifier is assigned, which consists of three parts separated by hyphens:

1. **Project identifier**—Assigned to the project when the project is initiated, this can be in a format chosen as appropriate for the particular enterprise.
2. **Requirement identifier**—Consists of three digits from 001 through 999, usually sequential within the project, which identify the requirement number.
3. **Revision identifier**—Consists of two digits which identify the requirement revision number. At the time when a requirement is approved and baselined, those

Table 2.1 Requirements Template

Project Identifier:			Project Name:			
Project Manager:			**Client:**		**Date:**	
Req. #	**Requirement Unique Identifier**	**Requirement Group**	**Functionality or Explanation**	**Rationale for Requirement**	**Priority**	**Impact Analysis Results**
		☐ Functional ☐ Nonfunct. ☐ Bus. level			☐ Must have ☐ Should have ☐ Nice to have	
		☐ Functional ☐ Nonfunct. ☐ Bus. level			☐ Must have ☐ Should have ☐ Nice to have	
		☐ Functional ☐ Nonfunct. ☐ Bus. level			☐ Must have ☐ Should have ☐ Nice to have	

two digits are always 00. Each subsequent approved change request will increase this number by one.

Examples of a unique identifier are *CLI00253-001-02* and *RET00229-011-00.* This identifier will be used throughout the life of the project to ensure that all project scope changes are identified. Documentation concerning each requirement includes its unique identifier, a description of the functionality to be provided for a functional requirement, the rationale for the requirement which states why the requirement exists (from a business perspective), priority and an impact analysis of the requirement on the business, the other requirements and existing products.

The requirements will be documented in accordance with the established requirements template. Additionally, a traceability matrix will be used to document and manage requirements in order to assist in traceability. The requirements document and traceability matrix must be stored in the project control book.

Requirements Template

The requirements template is shown in Table 2.1. Explanation of requirements group and priority is provided in Chapter 3 in the discussion of processes R4-1 and R4-2.

Perform/Update Cost-Benefit Analysis (R3a and R3b)

The project cost-benefit analysis is a documented process which is used in decision making to determine whether the project is worth the money spent on its development. Benefits received from the project within a reasonable period of time must exceed the cost of the project. Process R3a is executed after the initial estimates are made, whereas process R3b is used after the preliminary estimates are made following completion of the business requirements document.

The benefits statement must be provided by the client when the initial project request is submitted. Since the client requests the project, it is assumed that the client believes that the project will be beneficial to the business. If the project is the development of a new marketable product (or service), then the marketing (or product/service management) department must have an answer to how many products or services may be sold and at what price. However, if the project is to redesign a section of highway to reduce accidents, the benefit cannot be immediately calculated.

The cost of a project comes from project estimates. Usually, the project manager receives hourly rates for project resources from the accounting department or from a delivery manager. It is important to note that hourly rates and hourly earnings are not the same. It should come as no surprise that if the resource cost is, say, $90 per hour, the salary component may only be around $45 per hour. Vacation, sick leave, pension, health benefits, the cost of facilities or equipment used for development, the overhead cost of management, marketing, etc. must be added to present the *true* cost of resources.

The cost of a project consists of several types of costs:

$$\text{Project cost} = \text{Direct cost} + \text{Indirect cost} + \text{Fixed cost} + \text{Variable cost} + \text{Sunk cost}$$

Direct cost is the cost incurred due to the work performed on the specific project. It includes wages, bonuses, work travel, materials, etc.

Indirect cost is the cost of running the organization, which is split among all company projects, such as cost of facilities, power, accounting, security, etc.

Fixed cost is the cost that does not change with the size of the project or its duration. This is a nonrecurring cost. Examples of fixed cost include machinery setup cost, a one-time cost of advertising for a special resource and so on.

Variable cost is the cost which grows in direct relationship to the size and length of the project, such as cost of material, equipment amortization, etc.

Sunk cost is the cost which has already occurred and over which we no longer have any control. For example, if we spent $300,000 on a project and then found that due to the wrong approach we have to start all over again, that money represents sunk cost. Sunk cost is a loss, which should not play any role in determining the future of the project. Only new estimated costs of the project and risk must be considered in the decision to stop or continue the project. However, sunk cost is indeed a part of the overall project cost.

The following are major formulas used for cost-benefit analysis.

Net Profit after Taxes

$$\text{Profit} = \text{Revenue} - \text{Cost} - \text{Expense} - \text{Project cost}$$

Revenue is company income as a result of business initiated due to the project. *Cost* is the money spent to organize the project. *Expense* is what is spent to obtain benefits from the project. *Profit* is the value of the benefits less the money spent to get the benefits. Profit is also called net earnings before taxes (NEBT) or gross profit. *Net profit* is also called net earnings after taxes (NEAT).

Return on Investment (ROI)

$$\text{ROI} = \text{NEAT}/(\text{Cost} + \text{Expense} + \text{Project cost})$$

(Cost + Expense + Project cost) is called *total investment.*

Present Value

Let us assume that you want to buy a new car. One dealer advertises the car with a "buy now, pay later" deal for $20,000, with payment due 24 months from now. Another dealer advertises the same car for $19,000, due immediately. You know that the current interest rate is 5%. Which dealer (and which *deal*) would you choose?

Very rough calculation would show that depositing $20,000 in the bank for two years would yield over $2,000 in interest, so you decide to pay the first dealer $20,000 in 24 months and make $2,000 in interest. If you were to pay the second dealer $19,000, you would save only $1,000.

Today's money is worth more than the same amount later. *Present value* (PV) is today's value of future cash flows. For example, if the project cost today is $1,000,000, two years from now, when money for the project is received by the project developer, the value of the money may be worth $800,000 due to inflation, rising costs, etc. In this case, the project developer is losing $200,000, but the client, who pays for the project, gains $200,000. This is called *PV cost.* Similarly, if you are a buyer and paid cash for the project in advance, by the time the project is done two years later, the value of the cash paid will be less, because you will not get interest on the money paid and you will not get project revenues for two years. After you start receiving project benefits, this lost revenue must be accounted for in the overall revenue calculation. This is called *PV revenue.*

NOTE The term PV is also used in earned value analysis (planned value). It is important to be aware of the context of the terms so that you apply them properly, especially with commonly used acronyms such as PV.

Since we are doing the calculation today, we want to know today's value of money that would be paid or received t years in the future:

$$\text{PV} = \text{FV}/(1 + r)^t$$

PV is the present value of money, FV is the future value of money, r is the interest rate (also called value discount rate) and t is the time period in years.

Practically, revenues, expenses and project costs do not occur at one specific moment, but rather are distributed over time, which makes accurate calculations difficult. Considering the low accuracy of initial project estimates and the estimated revenues, for the sake of calculation, it can be safely assumed that all revenues and costs happen at one specific moment at the end of each year, and the cost-benefit analysis is calculated separately for each year. Refer to the example of cost-benefit analysis calculation provided in this section.

Cost-Benefit Ratio

The cost-benefit ratio (CBR) is the expected profitability of a project:

$$CBR = PV_{revenue}/PV_{costs}$$

A CBR of 1.0 is the breakeven project (no profit). A CBR <1.0 means costs exceed benefits (and means a project is not financially attractive). A CBR >1.0 means a profitable project.

Payback Period

The payback period is defined as the number of months up to the point where cumulative revenues exceed the cumulative costs and the project pays for itself:

$$\text{Payback period} = \text{Project cost}/[(\text{Project cost} + \text{NEAT}_{1st\ year})/12]$$

Note that if revenue for the first year is less than all expenses, NEAT will be negative.

If the project cost is $1,000,000 and NEAT in the first year after project completion is –$200,000, then:

$$\text{Payback period} = \$1{,}000{,}000/[(\$1{,}000{,}000 - \$200{,}000)/12] = 15 \text{ months}$$

When calculating cost-benefit analysis, it is very important to avoid double counting costs or benefits. For example, if the first-year cost-benefit analysis includes the cost of the project, it is not counted again in the second-year cost-benefit analysis. However, if the project does not pay for itself in the first year, the unpaid part is counted as the project cost of the second year. In this case, for the second-year calculation, the time period in years for the PV cost calculation part related to project cost is 1.

Similarly, if revenues are received each year, rather than at the end of a period of several years, then the time period in years for the PV revenue calculation is always 0, because costs and revenues are counted at the end of each *year.*

While the calculations are not complicated for a project manager to do, it may be a good idea to have an authorized person, such as a financial analyst, deliver the formal statement of expected payback period in months. The general guideline is that the expected payback should not exceed 24 months. If it is found that the project financially does not justify itself, the decision is made by the project manager to reject the initial project request. *It is wise to consult with your line management prior to notifying the client so as to be aware of other reasons which could take priority over financial considerations of the project (for example, compliance with regulations).* It must be clearly stated in the cost-benefit analysis document that despite the lack of financial benefits, the project request is approved due to reasons specified. The cost-benefit analysis document must be signed by the project manager and the development manager.

Another consideration in these analyses is the longer term sustainability and the so-called "triple bottom line" (people, planet, profit). These are international criteria for measuring the success of an organization in economic, ecological and social areas. While a country's laws supersede international laws, it may be appropriate in your role as project manager to ensure that your project is in alignment with corporate social responsibility or the corporate mission, which often helps dictate this long-term view.

Cost-Benefit Analysis Example

Let us assume the client submitted the following benefit analysis:

1. The proposed project will develop a website for booking vacation holidays. The website will allow customers to select, book and pay for vacations online.
2. After completion of the project, the expected rate of the online holiday bookings will be 500 vacations per month for the next three years or 6,000 vacations per year.
3. The average price of one booked vacation is $2,000.
4. The annual cost of operations is estimated to be $75,000.
5. The cost of project development is estimated at $1,500,000.

Refer to the cost-benefit example in Table 2.2. During each of the first three years of operation after the project ended:

- Vacations sold per year: 6,000 (line 1 = actuals by the end of each year)
- Unit price of each vacation: $2,000 (line 2 = actuals)
- Unit expenses (room maintenance): $1,700 (line 3 = actuals)
- Gross revenue: $12,000,000 (line 4 = calculated)
- Expenses (paid to hotels and airlines): $10,200,000 (line 5 = calculated)
- Revenue: $1,800,000 (line 6 = calculated)
- Cost of operations: $75,000 (line 7 = actuals)
- Project cost: $1,500,000, applied only to the first year (line 8 = actuals); since we do not want to count it more than once, the project cost for the second and third year is 0

PV cost in our example consists of PV project cost plus PV operating cost. Two separate calculations are required here, because the time period *t* is not the same for both. After these calculations, both PV values are added to get the PV for all costs.

Even though the operational costs occur throughout the year, the earlier assumption was to account for them at the end of the year, so that no time passed from the assumed operational costs payment and the end of the year. Thus, *t* for PV operating cost is 0 for all three years of operation. For the first year of the PV project cost calculation, *t* is

Table 2.2 Cost-Benefit Analysis Example

		Formula	1st year	2nd year	3rd year
1	Vacations sold per year (units sold)		6,000	6,000	6,000
2	Price for one vacation holiday (unit price)		$2,000	$2,000	$2,000
3	Expenses for one vacation holiday (unit expenses)		$1,700	$1,700	$1,700
4	Gross revenue	Units sold * Unit price	$12,000,000	$12,000,000	$12,000,000
5	Expenses	Units sold * Unit expenses	$10,200,000	$10,200,000	$10,200,000
6	Revenue	Gross revenue – Expenses	$1,800,000	$1,800,000	$1,800,000
7	Operating cost		$75,000	$75,000	$75,000
8	Project development cost		$1,500,000	$0	$0
9	Total investment	Project cost + Operating cost	$1,575,000	$75,000	$75,000
10	NEBT	Revenue – Operating cost – Project cost	$225,000	$1,725,000	$1,725,000
11	Tax rate		20%	20%	20%
12	Taxes	NEBT * Tax rate	$45,000	$345,000	$345,000
13	NEAT (revenue)	FV net earnings = NEBT – Taxes	$180,000	$1,380,000	$1,380,000
14	Return on investment	NEAT/Total investment	11.43%	1,840.00%	1,840.00%
15	Time period *t* (in years) from payment of project development cost		1	2	3
16	Time period *t* (in years) from payment of maintenance cost		0	0	0
17	Time period *t* (in years) from receiving annual revenues		0	0	0
18	Interest rate *r*		5%	5%	5%
19	Present value (PV project cost)	PV1 = FV project cost/$(1 + r)^t$	$1,428,571	$0	$0
20	Present value (PV operating cost)	PV2 = FV operating cost/$(1 + r)^t$	$75,000	$75,000	$75,000
21	Present value (PV all costs)	PV1 + PV2	$1,503,571	$75,000	$75,000
22	Present value (PV revenue)	FV net earnings/$(1 + r)^t$	$180,000	$1,380,000	$1,380,000
23	Cost-benefit ratio	$PV_{revenue}/PV_{all\ costs}$	0.12	18.40	18.40
24	Payback period (months)	Project cost/[(Project cost + NEAT)/12]			10.7

1, because the payment occurred one year ago. For the second and third years, t is 2 and 3, respectively. Once all PVs are calculated, they are added to arrive at PV all costs.

Similarly, the time period t for PV revenue is 0 for all three years, since revenue is assumed to occur at the end of each year. Since t for PV revenue is always 0, you may wonder why we bother making this calculation. Actually, this is not always the case. For example, in construction projects the future occupants often pay a deposit in advance, while the building is being built. In this case, t will not always be 0.

The payback period calculation result is 10.7 months in this example, which means that the project cost will be paid back in 10.7 months, after which only operating costs and unit expenses will constitute project costs, resulting in excellent profits beginning in the second year of this business.

However, if the payback period is over 12 months, then NEAT for the first year will be negative, since expenses in the first year, which include the cost of the project, exceed earnings. In this case we have to figure out first the payback amount in the first year and then use the balance of project cost minus payback amount for the first year as the project cost for the second year. The following formula can be used to calculate the balance for a payback period over 12 months:

$$\text{Balance} = \text{Project cost} - [(\text{Project cost}/\text{Payback period}) * 12]$$

For example, if the calculated payback period is 18 months, then two-thirds of the project cost ($1,000,000) will be paid back during the first year and the rest, $500,000, during the second year:

$$\text{Balance} = 1{,}500{,}000 - [(1{,}500{,}000/18) * 12] = 500{,}000$$

This amount, $500,000, is entered in the second-year column on line 8.

Create Project Control Book (R2)

The project control book (PCB) is a tool for storage of all project documentation. The tool is set up in the Requirements Frame and extensively used throughout all project frames. The purpose of this section is to define required content for the PCB as well as to define methods of classification and documentation of all project-related events in a way which allows efficient and straightforward access to the information stored in the PCB.

The overall guideline for PCB content is to store absolutely everything related to a project, because even one small omission may cause misunderstanding and lead to serious project consequences.

The following is one example of overall PCB content:

- Project standards, practices and methods
- Agreements and contracts
- Project deliverables and milestones

- Project delivery team members' information
- Plans
  - Project schedule
  - Communication management plan
  - Project risk plan
  - Quality assurance plan
  - Configuration management plan
  - Project training plan
  - Staffing plan
  - Other plans as created
- Meeting minutes
- Project status reports
- Project scope changes
- Risk assessment
- Project estimates
- Project issues and issue tracking
- Project financials and tracking information
- Metrics
- Approvals
- Project tools
- Quality assurance reports
- Contents of e-mails
- Contents of verbal communications
- Other project documentation

For the Requirements Frame, the following should be documented, as a minimum:

- Initial project request and benefit statement
- Cost-benefit analysis
- Initial project estimates
- Requirements Frame plans and schedule
- Meeting minutes with client to define project business requirements document
- Issues
- Approved business requirements document
- Traceability matrix
- Project charter
- Scope change requests
- Project status reports
- Quality assurance reports

In the frames which follow the Requirements Frame, there will be more documentation to file in the PCB. With such a large amount of information, it will be increasingly difficult to find a required document without structured electronic document storage.

The PCB should be easily accessible in a secure shared location, so that all key project stakeholders can see project documentation. However, there may be some documents that have restricted access. For example, clients should not normally see minutes from the internal project team meetings or the detailed project schedule. Team members are not supposed to see other members' information, etc. When the folder structures or documents are created, access authority must be thoughtfully established. Your enterprise's security administrator can set the access rights to folders or documents.

There is absolutely no justification for not including any project event or document in the PCB. Even the project issues brought up in casual conversation with clients should be documented in the PCB.

Off-the-shelf document management packages are available, which may be used to help implement the PCB, but the simplest way to build the PCB is to use the MS Windows file structure. The top-level folder is "project name," with further breakdown by project frames, by months, etc. Within the Requirements Frame there will be other folders, such as the traceability matrix, project plans, etc. The project plans folder will have subfolders such as project schedule, communication plan, quality assurance plan, risk management plan, staffing plan, etc. It is preferable to have the date as the first part of the document name within the structure indicated here, such as "2013-01-12 Team Status Meeting" or "2013-02-03 Risk Assessment," in order to sort all documents in each folder by date. This PCB structure allows easy access to and easy retrieval of documents.

The entire PCB structure does not have to be built right at the beginning of the project. It will be elaborated as the project progresses.

3

Business Requirements Analysis (R4)

The Business Requirements Analysis (R4) process starts when the project flow returns from the Planning Frame after the business requirements analysis has been planned (process entry point 2A). The Business Requirements Analysis is decomposed into three elementary processes, as shown in Figure 3.1:

1. Elicit Detailed Requirements (R4-1)
2. Assign Requirements Priority (R4-2)
3. Establish Acceptance Criteria (R4-3)

Elicit Detailed Requirements (R4-1)

Requirements Gathering Techniques

A requirement is a condition, functionality or capability of a product or service needed by a user to solve a business problem. It must be related to the needs of the business, rather than to technology constraints. If a business specifies that the new system must use an Oracle database, this is a technology constraint rather than a business requirement. Even though this may be taken into consideration, if it makes sense from a technical standpoint, the business requirements document should not include it as a business requirement. Requirements answer the following questions regarding the state of the business and applicability to the project: Who? What? Where? When? How much? *Requirements must not state how the problem should be solved in terms of technology.*

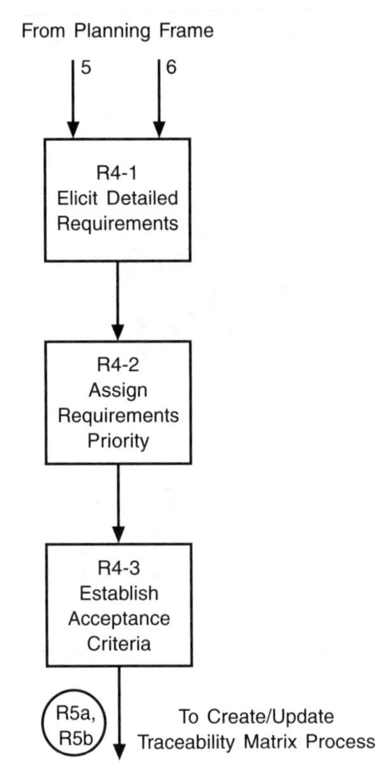

Figure 3.1 Business Requirements Analysis (R4) Process Flow

There are a variety of techniques available for requirements gathering. The most common are:

1. Interviews
2. Formal methodologies
3. Joint Application Design (JAD) sessions for certain types of projects
4. Brainstorming
5. Research and study of existing solutions
6. Surveys and questionnaires
7. Observation of user's work
8. Prototyping (for electronics, machinery and software projects)

Interviews

There are specific interviewing methods which allow determination of the current and desired business processes, as well as help to decompose large business functions into smaller elements.

When the project is an implementation of the enterprise business strategy, the interviewing process must start with the senior-level executives to obtain their point of view on the required business strategies. Interviews continue with business users at the lower management level and finally with those who are most familiar with different elements of the business throughout the enterprise.

The purpose of interviews is to gather as much detailed business information as possible, especially from those closest to the work. Based on interviews, main business functions may be identified and decomposed into smaller processes, usually building process decomposition and process flow diagrams.

Each element of decomposition must be verified by users from different business areas. Each business process should be decomposed to the level of the elementary process, which is a process that cannot be decomposed any further without losing business meaning. Thus, the business process Print Invoice may be decomposed into Print Header, Print Line, Send End-of-Line, Send Carriage Return, etc., but none of these processes has any business meaning. This means that the Print Invoice process is an elementary process and decomposition must be stopped there. Process decomposition and process flow diagrams are shown in Figure 3.2. The upper two drawings are process decompo-

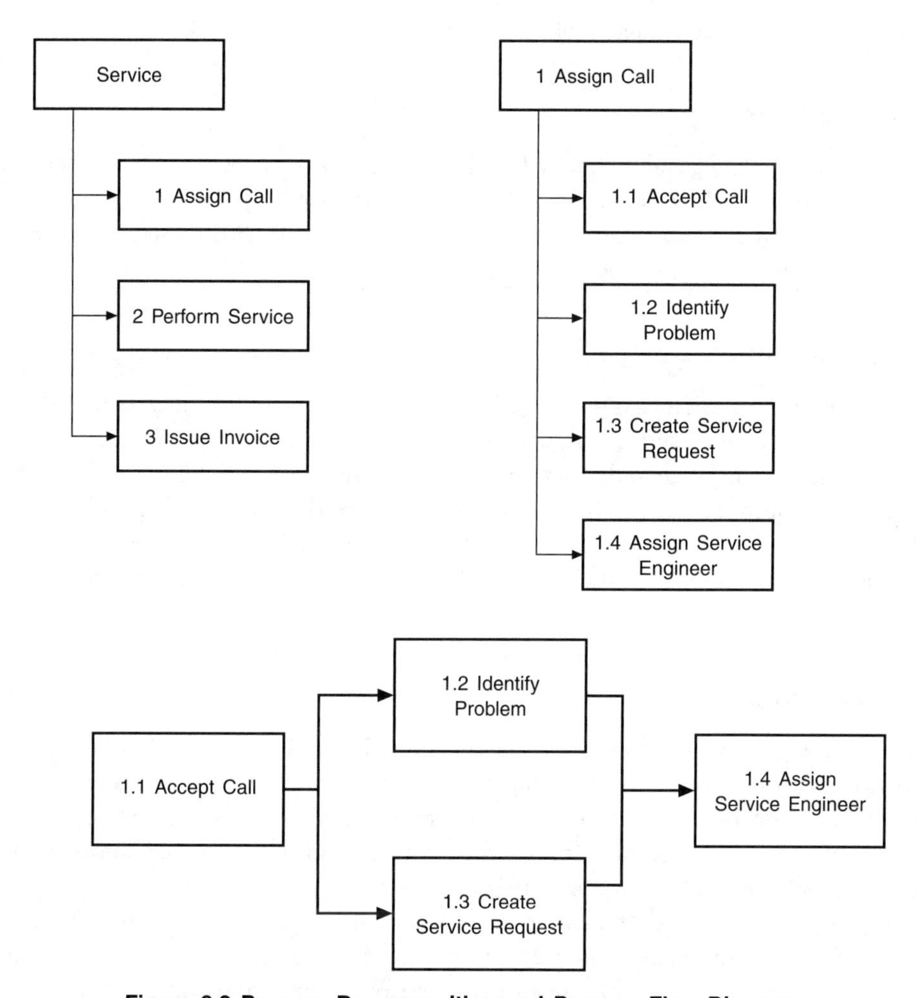

Figure 3.2 Process Decomposition and Process Flow Diagram

sition diagrams for a service provider's business. The top left drawing displays decomposition of a service business. It breaks down into high-level business functions:

1. Assign Call
2. Perform Service
3. Issue Invoice

The Assign Call business function may be further decomposed into:

1. Accept Call
2. Identify Problem
3. Create Service Request
4. Assign Service Engineer

The drawing at the bottom of Figure 3.2 displays the process flow diagram for the Assign Call business function. It shows, in addition to all elements of the Assign Call decomposition, the dependency of its elementary processes.

UML Methodology

The popular modeling methodology for object-oriented software development projects is the Unified Modeling Language (UML) and tools based on UML, such as the IBM Rational Unified Process (RUP) tool and others. The use of UML-based tools requires special training.

Joint Application Design

Joint Application Design (JAD) is a methodology that involves business participation in two- or three-day workshops together with delivery team members. JAD may be useful if a large number of stakeholders are present with many areas of interest and sometimes competing requirements. The JAD session must follow the formal JAD process and the JAD structure in its entirety. Otherwise, it will not achieve its objectives.

The JAD moderator role, which guides the workshop sessions to be an effective tool, requires special training.

Brainstorming

Brainstorming sessions are used to generate, combine and further develop good ideas, rather than propose solutions. Facilitating a brainstorming session requires the skill and experience to encourage participants in generating ideas, to avoid one person's domination and to extract the best, most creative ideas and at the same time to prevent participants from lengthy, nonproductive discussion.

In a requirements management brainstorming session, participants should share their vision of each requirement presented and expand it. Once the session is documented, it is necessary for the project manager to send session minutes to participants and request their confirmation that the minutes—and, more importantly, the ideas—were recorded correctly.

Research and Study of Existing Solutions

Research of the published documentation in the company and on the Internet may provide insight into requirements and functionality of similar systems. This may help your team avoid reinventing the wheel.

Observation of User's Work

Observation of the user's work will help in understanding the current business working environment and in understanding the commensurate requirements in context.

Surveys and Questionnaires

Surveys of existing business practices and questionnaires on this subject will help to understand the current environment. Surveys and questionnaires can be developed when requirements are understood well in order to confirm compatibility of existing business practices with new requirements. The vast majority of business users will fiercely resist changes to their existing business practices, because it is simply human nature to resist change. Changes should therefore have the strong backing of both senior business users and clients. Consequences of not taking this advice will be business complaints to management and even requests to replace the project manager.

Prototyping

This method is used for software, electronics and mechanical design projects. There are several levels of prototyping, from drawings on paper to developing functional prototypes. The Agile methodology for software development may be used for functional prototypes. When building application screens, business users may propose layouts and screens, but the final screen layout should come from usability experts or from experienced technical personnel.

Requirements Criteria

In order to succeed in using various requirements gathering methods and tools, the requirements manager must be trained in this area.

All requirements must satisfy the following criteria:

- **Necessary**
  - ◇ The requirement is essential for the product to perform its function.
  - ◇ If the requirement is removed, the product will not be able to function as required.

- **Concise**
  - ◇ There is one requirement per requirement statement.
  - ◇ The requirement clearly states what must be done.
  - ◇ The requirement avoids nonessential information.
  - ◇ The requirement is easy to understand.
- **Construction/Design Free**
  - ◇ The requirement states what is required, not how to implement it.
  - ◇ The requirement is easy to understand.
- **Achievable**
  - ◇ The requirement is technically feasible to implement as confirmed by subject matter experts.
- **Complete**
  - ◇ The requirement is complete and no clarifications are needed.
- **Consistent**
  - ◇ The requirement does not cancel or undermine other requirements.
  - ◇ The requirement is consistent with terminology throughout the business requirements document.
  - ◇ The requirement is stated only once in the business requirements document.
- **Unambiguous**
  - ◇ The requirement should have a single accurate interpretation without ambiguity by all project stakeholders.
  - ◇ The requirement should not use technical terminology or buzzwords and must be understood by clients with no technical knowledge.
  - ◇ The requirement must avoid using ambiguous words, such as efficient, user friendly, easy, fast response time, etc., without quantification.
- **Verifiable**—The implemented requirement, which will be translated into a deliverable, must be verifiable by (in order of verification accuracy) inspection, analysis, demonstration and test of the deliverable:
  - ◇ *Inspection* is a visual verification of the specified criteria and parameters. Often this method alone is not sufficient.
  - ◇ *Analysis* is the comparison of parameters against pass/fail criteria.
  - ◇ *Demonstration* is the activity which proves to the client that the product of the project conforms to requirements.
  - ◇ *Test* is physical measurements in the real or a simulated production work environment.
- **Traceable**—Each requirement that conforms to all the above conditions has a unique identifier in order to trace all changes to it throughout the project life cycle.

A requirement is the intent to have a specific deliverable. Every requirement must point to a real-life specific deliverable, which must be unique, must exist in the real world and must be measurable:

- **Unique**—There is no other deliverable which results from other requirements that is identical to this deliverable.

- **Exists in the real world**—You can actually see and interact with the deliverable. You also should be able to determine whether the deliverable is compliant with the requirement.
- **Measurable**—Its characteristics can be measured. If, for example, the requirement says a device must operate in a temperature range of 0 to 95°F, the actual deliverable must be subjected to the above range of temperatures to confirm that it conforms to the requirement.

Examples of requirements are:

1. Produce a weekly financial report
2. Develop an independent suspension for a vehicle
3. Produce a business requirements document

The corresponding deliverables for the above requirements are:

1. The actual weekly report (electronic or on paper)
2. Mechanical assembly which is part of a vehicle
3. Business requirements document

Requirements must be technology independent. However, selecting a requirements gathering methodology is specific to the tools and technology used. Tool selection depends on the technology used. For example, if the project deliverable is a web-enabled software application, there is a good chance that the tool used will support the UML, like RUP. RUP has specific sets of requirements gathering and documenting techniques, such as use case, interaction diagrams and others. If the project deliverable is intended to be a software mainframe application, mechanical design or a workflow, RUP is not suitable.

Requirements Groups

All requirements may be broken down into three groups:

1. Functional requirements
2. Nonfunctional requirements
3. Business-level requirements

Functional requirements focus on what the system *does* in the sense of discrete functionality of the deliverable. The following are examples of functional requirements:

- The automobile must have a five-speed manual transmission
- The application must present an employee's personal information on the screen
- The building must have ten-inch brick walls

Nonfunctional requirements are usually needed to establish parameters of product operation. They may be related to the environment, administration, maintainability, documentation, training, reliability, disaster recovery, etc. Examples of nonfunctional requirements include:

- The system will operate in the temperature range between 0 and 120°F
- The mean time between system failures is 5,000 hours
- Only authorized users will be able to access their bank account information online
- In case of failure, the production line must be repaired in less than one hour
- Three hours of training must be provided to all users

Business-level requirements are requirements that do not fall under the above categories, but rather are related to a temporary or permanent business strategy, tactic or specific regulation. Examples of these requirements are:

- Government or organization-regulated requirements, like Sarbanes-Oxley, tax code, etc.
- Strategic requirements, like a goal to increase revenue by 10%
- Tactical requirements, like a temporary discounted interest rate for new credit card customers

Example of Requirements Clarification

Let us assume that the two following business requirements in the banking area were presented by the client:

1. Allow an overdraft limit of up to $500 for checking accounts
2. Allow fast search by customer name

The above requirements will be analyzed and the following clarifications will be requested:

1. Are there any entitlement conditions to allow an overdraft, like having a savings account with sufficient funds, a mortgage, etc.?
2. Suppose a customer has $100 in an account and intends to withdraw $650 using an ATM. Should the customer be offered $600 instead by a screen message or is the transaction declined altogether?
3. What does "fast search" mean? Is three minutes fast enough? Exactly what does search by "customer's name" mean? Must the teller enter the *complete* first and last names? Should the search be allowed only by last name? Should wild card characters like "Sm*" be allowed to retrieve all last names starting with the letters "Sm"? Should substitution characters be used, like %son? Should the first name be included in the search for partial names entered? Should there be additional search conditions, like birth date, address, etc.?

Once requirements are clarified during interviews with clients, the requirements manager also should get the opinion of the qualified delivery team members to confirm that the requirements are feasible to implement and will not cause degradation of system performance, especially when exccuting a search. What would be the feasible quantifiable response time for a search? After modifications, the above business requirements may read:

1. Checking accounts should be allowed an overdraft limit of up to $500, provided the customer has a savings account with total savings that exceed the overdraft amount. If the withdrawal amount requested by the customer exceeds funds in the account plus $500, the transaction must be declined altogether.
2. An authorized bank employee will be allowed to perform a customer search by entering the full last name in the last name field on the screen. In order to limit the list of customers retrieved, the bank employee may also enter one or more alpha characters (A to Z) of the first name in the first name field and/or numeric characters and "/" in the date of birth field for the full date of birth in the form MM/DD/YYYY. The response time (defined as the time from the moment the enter key is pressed to initiate the search until results of the search are displayed on the screen) will not exceed five seconds.

Assign Requirements Priority (R4-2)

When there are multiple requirements, the requirements manager may assign requirements priority. The customer and affected groups participate in the prioritization. This prioritization activity should be documented in the project control book, including the date and attendees, the results of the prioritization, changes that occur and approval of changes from the requirements source. There are three levels of requirements priority:

- Must have
- Should have
- Nice to have

A "must have" requirement is a requirement without which the application is not functional. A "should have" requirement is a requirement which is needed to achieve a business goal, but if not implemented the application may temporarily function without it. A "nice to have" requirement is the bells and whistles. An application may function indefinitely without it. Although beyond the scope of this book, we recommend the use of Kano analysis in determining the "story behind the story" in identifying and prioritizing customer requirements.

Prioritization is very important from a practical point of view. Regardless of how thorough an analysis is, clients will inevitably change their minds many times later in the project and issue a multitude of scope change requests. Clients often forget that each change request costs money and the project budget may be exhausted long before the end of the project. The additional budget comes reluctantly and is usually insufficient to cover all scope changes. At this point it is very useful to start cutting down first on "nice to have" requirements and then even on some "should have" ones. In order to let clients save face, the project manager should suggest temporarily postponing implementation of the "nice to have" and even some "should have" requirements until the new budget is available. In the real world, money almost never comes later, because other budget priorities will tend to take priority over "nice to have" requirements.

Establish Acceptance Criteria (R4-3)

The acceptance criteria for the project and each requirement must be included in the requirements document. Acceptance criteria are the *relevant, specific and measurable* criteria that requirements must be subjected to in order to be accepted by the client. This is not what clients would like to see, but rather what has been agreed to earlier and documented. The acceptance criteria will be used to develop test cases for acceptance testing in the Closing Frame of the project. For example, let's say the requirement for a wind power generator says it must produce a minimum of 10 kW of power at winds of 2 miles per hour or stronger:

- The requirement is *relevant,* because producing energy is the objective of the wind generator.
- The requirement is *specific,* because it quantifies the energy produced (10 kW of power).
- The requirement is *measurable,* because there are practical methods to simulate the wind and to measure the power output when the wind power generator is manufactured.

However, if a requirement says that the wind power generator must be energy efficient, that is not specific and therefore cannot be measured. Therefore, it is impossible to determine the relevant, specific and measurable acceptance criteria for this requirement.

Let us look at the following requirement for the same wind power generator. The wind power generator's life span must be no less than 5 years at average winds of 2 miles per hour and sustained temperatures up to 90°F. How do you measure this? You can create a simulated environment based on an engineering calculation. If engineers confirm that winds of 100 miles per hour at a temperature of 140°F increase the wear on the generator by 60, then running the wind power generator for 1 month will be roughly equivalent to 5 years at the nominal environment.

Requirements Management

Create/Update Traceability Matrix (R5a and R5b)

Requirements traceability is defined as the ability to describe and follow the life of a requirement throughout the entire project life cycle. Timely updates to the traceability matrix, along with adherence to the project scope change process, will avoid undocumented scope changes and adhere to the approved project scope. The traceability matrix will provide the necessary documentation for project quality assurance reviews later in the life cycle.

The traceability matrix can be created by placing the uniquely identified requirements in the matrix. This information must be captured in the project control book and tracked to accomplish traceability. The matrix is used throughout the life of the project to assist in impact analysis and implementation of project changes to all the project work products and deliverables.

When the Business Requirements Analysis (R4) process is completed the very first time, the traceability matrix does not exist yet and the process flow goes to the Create Traceability Matrix (R5a) process. In process R5a, the traceability matrix is created from the business requirements documented in Table 2.1, using the traceability matrix template, which lists each uniquely identified requirement in a matrix. A reference to the corresponding change request is provided.

In the case where the Business Requirements Analysis (R4) process is conducted in order to analyze a scope change request, the process flow is directed to the Update Traceability Matrix (R5b) process. As the project progresses, each new change request is updated in the matrix by adding an identifier with a new version of the requirement, date, frame and description of the change. The traceability matrix is a dynamic document, which is appended each time a project scope change request is approved. Table 4.1 is an example of a traceability matrix (the template is an Excel-based document), and Table 4.2 shows field descriptions.

Create Business Requirements Document (R6)

Once the new traceability matrix is created and all requirements are documented, the business requirements document (BRD) will be created. When signed off on by the requirements manager and client's management, this document represents a baseline for requirements. Project authorization is based on those requirements. The BRD is a reference document, which is never updated after its approval.

The BRD must have the following sections, at a minimum:

1. Glossary
2. Project scope
3. Overall business description
4. Project estimated price to client
5. Cost-benefit analysis
6. Requirements gathering methods
7. Requirements section
8. Major assumptions and constraints
9. Scope change management
10. Risk assessment
11. Evidence of approval

Usually, clients are not provided with project cost estimates. Rather, they are given the *project estimated price,* which is based on project cost estimates. This includes costs related to risk assessment plus a profit margin and other costs defined by the delivery organization. The project manager usually provides the estimated project costs to the cost analyst or to management and receives the price. The price must indicate the method used to produce estimates and their accuracy.

The BRD must provide the following information for each requirement:

1. **Requirement ID**
2. **Requirement description**—A description of the delivered functionality, with details and specifics.
3. **Rationale for the requirement**—The reason why the requirement exists and what it is expected to achieve, as well as its priority. If some requirements cannot be implemented due to lack of time or budget, priority will be used to identify requirements for cancellation or deferral.
4. **Requirement type**—Functional, nonfunctional or business-level requirement.
5. **Requirement priority**—Must have, should have or nice to have.
6. **Completion criteria**—Description of the process to confirm that the requirement is met.

Table 4.1 Traceability Matrix Example

Project ID:	*CLI00253*						
Project Name:	*Requirements Management Process Development*						
Project Manager:	*John Smith*						
Requirements Manager:	*Jane Smith*						
Requirement Initial Identifier:	*CLI00253-001*		**Priority:**	*Must have*	**Date:**		
Requirement Description:							
New Requirement Identifier	**Change Request ID**	**Date of Change Request Approval**	**Date Change Implemented**	**Requested by**	**Life Cycle Frame**	**Requirement Type**	**Description of Change**
CLI00253-001-01							
CLI00253-001-02							
CLI00253-001-03							
Requirement Initial Identifier:	*CLI00253-002*		**Priority:**		**Date:**		

Table 4.2 Traceability Matrix Field Descriptions

Field	Description
Requirement initial identifier	The initial identifier assigned to a requirement
Priority	Requirement priority ("must have," "should have," "nice to have")
Requirement description	Brief requirement description (detailed description may be found in the BRD)
New requirement identifier	Increase the requirement version component identifier by 1
Change request ID	The change request identifier that affects this requirement
Date of change request approval	Date when change request was approved (may also be due to a defect found)
Date implemented	Date when change request was completed
Requested by	Name of the person who submitted the change request
Life cycle frame	Life cycle frame in which the requirement change occurs
Requirement type	Three-character identifier: **FNC** for functional requirement **NFR** for nonfunctional requirement **BUS** for business-level requirement
Description of change	Brief description of the change required and reference/link to the change request

Major assumptions and constraints must indicate your understanding of the work environment. There must be a statement which says that if an assumption is not true, a new project estimate must be made, which may modify the project cost, duration or both. Also, we advise that you think of assumptions as a source of project risk. Any of them may escalate to a risk at any time. For example, you may assume that the exchange rate between two key countries in your project's scope is stable when in fact there may be a vast change in the exchange rate at any given time.

The following are examples of assumptions and constraints:

- The client and the required business representatives will be available for all scheduled weekly project meetings for not less than one hour each.
- All questions presented by the project manager will be answered by the client in writing within three business days.
- The third-party supplier will provide the equipment, as listed in the statement of work, no later than September 1, 2014.
- The developing organization will hire a Swahili language specialist no later than December 1, 2014.

The completed BRD, even if it is not signed off on yet, provides details about the project, which allow you to do the improved project estimates. The project flow goes back to the Planning Frame for the improved project estimates and then returns to the Perform/Update Cost-Benefit Analysis (R3b) process, unless it was already done in process R3a. If process R3a was done earlier, the cost-benefit analysis will be updated before executing the Conduct Requirements Review (R7) process.

Review Requirements

Requirements Self-Assessment Checklist

After gathering and documenting requirements, but before the formal review, the requirements manager must complete a self-assessment checklist. The checklist consists of two parts:

1. Checklist #1 is related to the entire Business Requirements Analysis (R4) process. There will be one checklist #1 for review.
2. Checklist #2 is related to one single business requirement, and there will be as many completed checklists as there are business requirements.

Both checklists have questions, which must be answered yes or no. If the answer is yes, there must be documented proof of the answer, such as meeting records, memos, documents, e-mails, the completed and documented elements of the project plan, a quality review report and protocols documented in the project control book. If proof does not exist, the answer must be recorded as no. If the answer to a question is no, then there is a gap, which must be closed before the formal review. When presented for review, all questions on both checklists must be answered yes.

Checklist #1 is displayed in Table 4.3. The template for checklist #2 is shown in Table 4.4. In addition to checklist #1, a separate checklist #2 must be filled in for each requirement. All questions must be answered yes to ensure that the requirement is necessary, concise, construction free, attainable, complete, consistent, unambiguous, verifiable and traceable.

Conduct Requirements Review (R7)

The Conduct Requirements Review (R7) process consists of the following elementary processes:

1. Identify Review Team (R7-1)
2. Schedule Review and Invite Participants (R7-2)
3. Send Materials to Participants (R7-3)
4. Conduct Review/Take Notes (R7-4)
5. Update Materials (R7-5)

The Conduct Requirements Review (R7) process flow is shown in Figure 4.1.

The purpose of the Conduct Requirements Review (R7) process is to review the documented requirements with the lead client and business users to ensure that these

Table 4.3 Checklist #1 Template

Project Requirements Review		
Project Name: ______ Date of Review: ______		
	Question	Yes/No
1	Do the project manager and delivery team members understand and comply with all company policies and procedures relevant to the project?	
2	Do the project manager and delivery team members understand the project scope?	
3	Is there a formal BRD and does it include requirements from all clients and stakeholders that participate in this project?	
4	Have all clients and the related business representatives participated in requirements gathering sessions?	
5	Is there a documented Requirements Frame plan package, which has been reviewed with the delivery team members?	
6	Is weekly project tracking being performed to determine the actual cost and schedule versus planned?	
7	Does the team for analyzing requirements and documenting them have sufficient experience in that business area?	
8	Was sufficient funding provided for the Requirements Frame?	
9	Are the delivery team members and requirements manager trained in methods used to perform their requirements analysis activities?	
10	Are the activities for analyzing the gathered requirements reviewed with delivery and business management on a periodic basis?	
11	Are the activities for analyzing the gathered requirements reviewed with the project manager on both a periodic and event-driven basis?	
12	Has the risk assessment been performed?	

requirements are complete and ready for use in project planning and for formal approval. Requirements are reviewed to ensure that they are necessary, concise, design free, attainable, complete, verifiable and traceable.

Identify Review Team (R7-1)

The review team should include the requirements manager, quality assurance analyst, project manager, relevant delivery team members, lead client and leading members of the affected business groups. The objectives of the review are to identify incomplete and missing requirements and verify that the allocated requirements are documented clearly and properly and are feasible to implement and test. All requirements will be reviewed with the team, the identified deficiencies corrected and the requirements document will be updated.

Table 4.4 Checklist #2 Template

Requirement Review		
Project Name: ______ Requirement ID: ______		
Requirement Description:		
	Question	**Yes/No**
1	Is the requirement detailed enough and adequate to develop an estimate and clear enough so the business knows what will be delivered?	
2	Is the stated requirement technically achievable?	
3	Is the stated requirement complete, with no further clarification required?	
4	Has it been verified that the stated requirement does not have a negative impact on other requirements and systems?	
5	Is the stated requirement unique in that it does not duplicate other requirements?	
6	Is the same terminology used for the same elements as in all other requirements?	
7	Does the stated requirement have one and only one interpretation?	
8	Is the language clear enough to leave no doubt about the intended descriptive or numeric value?	
9	Does the stated requirement use commonly used nontechnical terminology and avoid technical terminology?	
10	Does the stated requirement use standard words or phrases such as "shall" and "will" and avoid ambiguous words like user friendly, flexible, fault tolerant, state-of-the-art, simple, efficient, easy and minimum/maximum without precise quantification?	
11	Can the stated requirement be verified by inspection, analysis, demonstration or test?	
12	Has the stated requirement been assigned a unique identifier in order to follow its life through all frames of the project?	
13	Does the stated requirement have only the essential capability, only the essential physical characteristic, only the essential quality and nothing beyond the essential?	
14	Is the stated requirement clear, easy to understand and states only what must be done?	
15	Does the stated requirement document WHAT is required, not HOW the requirement should be met?	
16	Is the requirement traceable?	

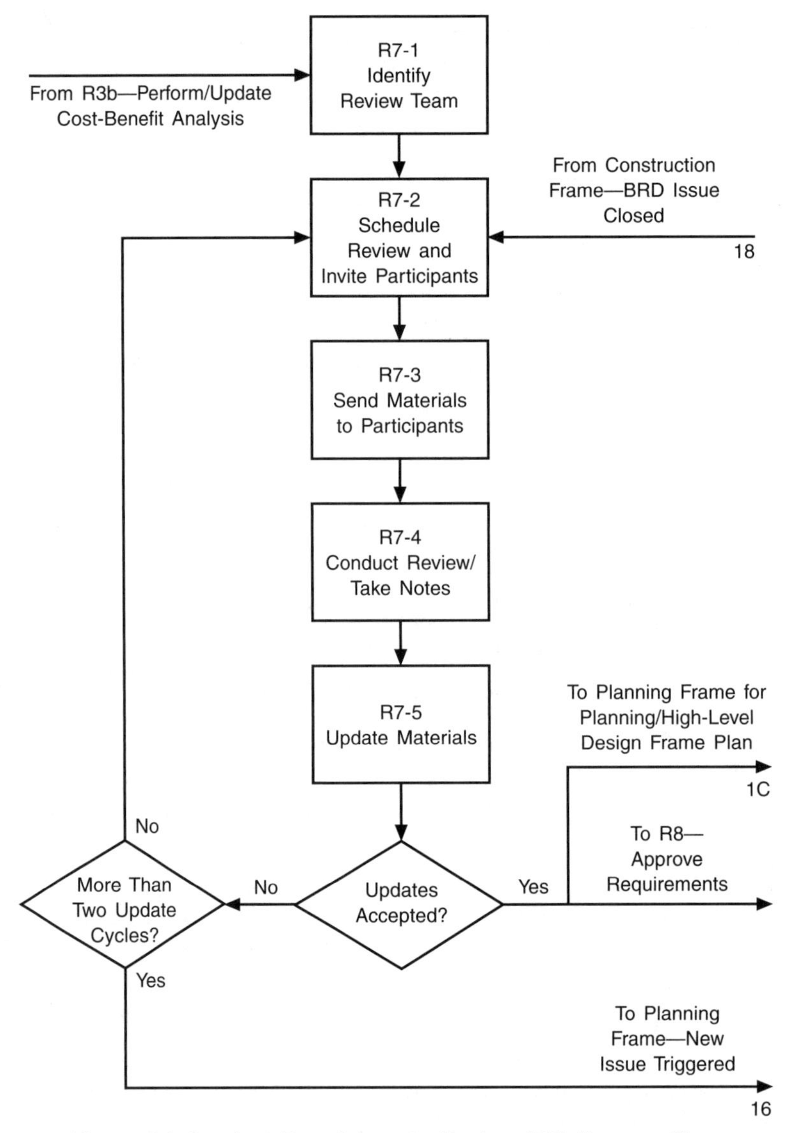

Figure 4.1 Conduct Requirements Review (R7) Process Flow

Once all the review team members are identified, a requirements review board list (Table 4.5) should be created. It is useful to enter names of business experts for each requirement in order to get answers quickly and efficiently and to be able to quickly address any question which comes up at any time during the course of the project. Make sure the lead client provides consent to access business experts directly.

Schedule Review and Invite Participants (R7-2)

The review must be scheduled at least seven days in advance, with the conference room booked for a face-to-face review or a teleconference set up. The meeting invitation and

Table 4.5 Requirements Review Board Example

#	Name	Role	Area of Expertise	Phone	e-mail
1	J. Smith	Client Lead	Overall Business	Ext. 2000	
2	A. Perry	Business Expert	Req. SA00234, SA00238	Ext. 2012	
3	N. Owen	Technical Specialist	Database Design	Ext. 1812	
4	F. Norris	Project Manager	Project Management	Ext. 1948	
5	———	———	———	———	

agenda will be sent to all participants. If any participant declines the invitation, he or she must ensure that a qualified representative is assigned to attend the review. The manager of that participant and the lead client must be immediately notified. If the requirements manager, the client lead, or business experts from critical areas are unable to attend, the review must be rescheduled.

Send Materials to Participants (R7-3)

All documents for the review must be sent to participants by electronic or regular mail and received by participants at least one week in advance. Materials include the BRD, completed checklists and other relevant documentation.

Conduct Review/Take Notes (R7-4)

The quality of requirements cannot be compromised under any circumstances. If the quality of requirements is compromised, the quality of the entire project (and, of course, its deliverable) will be compromised as well. Quality is conformance to documented and approved requirements and specifications, and therefore it is not the same as "excellence" or "perfection." Quality is not a part of any project constraint, unlike cost, time and resources. If the project duration, budget or resources are not realistic to deliver all business requirements, then some requirements should be dropped or changed to reflect the new situation as mutually agreed upon with the business. Still, all deliverables must be produced with the quality required by the business. For example, if the cost of delivering 100% of banking data within 600 milliseconds without losing any data is too high, the requirement cannot be compromised to deliver only 80% of data within 600 milliseconds, losing the rest of it. Instead, a new standard must be negotiated with the business to increase the delivery time to 1 or 2 seconds, which is a totally legitimate agreed level of quality. Exceeding the documented business requirements is sometimes called "gold plating" and is neither desirable nor acceptable, because this will increase the cost of the project over the amount that the client has agreed to pay. This may have negative consequences for the project and the project manager. Also, introducing new business features without a proper change request produces scope creep, an always undesirable element in all projects. Scope creep will be discussed later in more detail.

The requirements manager will present the BRD and checklists to the review team. All questions in the checklists must have already been reviewed by the project manager and answered yes. During the review session, the checklist #1 walk-through will be performed once; then the requirements manager will read each requirement one by one and verify it with the checklist #2 walk-through, followed by the BRD walk-through.

When all required modifications and fixes are identified, the project manager must obtain confirmation from the review team by sending the modified requirements to all review team members. This ensures that the review notes are correctly recorded and only required modifications and fixes are made. The review activity should be documented in the project control book, including the date, attendees, agenda, results of the review and the required modifications.

Update Materials (R7-5)

The requirements manager must perform all of the required modifications in the BRD and send them to the project manager, quality assurance analyst and the lead client. If the quality assurance analyst and the lead client are satisfied with these changes, and no additional issues are found, they notify the project manager that the requirements are accepted on that level. In this case, the Approve Requirements (R8) process is initiated. At the same time, the detailed planning of the next set of activities must be initiated so it is completed before the end of the Requirements Frame. The next project activities are the detailed planning of the Planning/High-Level Design Frame, as well as the overall planning of the rest of the project. The timing is important, because directly after completion of the Requirements Frame, the existing resources must be utilized immediately by starting the implementation and tracking of the high-level design. You may recall that planning of the high-level design was completed at the same time as the Requirements Frame was completed. In addition, the overall project plan to the end of the project was finished as well. The plan cannot be *too* detailed, because at that point only the completed requirements are available, which is not enough to plan and execute a project. However, having that plan allows doing the preliminary estimates of the project cost and schedule, which are required to obtain project authorization from the sponsor beyond the Requirements Frame.

If updates are not accepted, a new review is scheduled and the entire process is repeated. If updates are not accepted after two attempts, a new issue is triggered and the process flow is directed to the Planning Frame via exit point 16 for the issue resolution plan. The issue must be resolved in accordance with the issue management process described in Chapter 8 (in Section III on the Planning Frame). When the issue is resolved, the flow returns from the Construction Frame, where it was tracked, via entry point 18, and the process is repeated.

Approve Requirements (R8)

The Approve Requirements (R8) process is displayed in Figure 4.2. By the time this process starts, the lead client and the business should have agreed that the BRD is

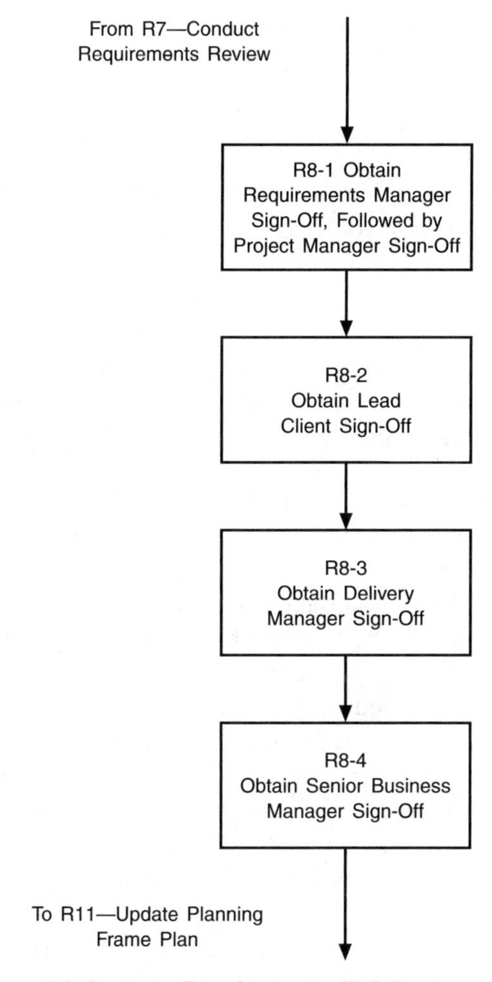

Figure 4.2 Approve Requirements (R8) Process Flow

complete. At this point, the formal sign-off process will be kicked off. This activity consists of the following processes:

1. Obtain Requirements Manager Sign-Off, Followed by Project Manager Sign-Off (R8-1)
2. Obtain Lead Client Sign-Off (R8-2)
3. Obtain Delivery Manager Sign-Off (R8-3)
4. Obtain Senior Business Manager Sign-Off (R8-4)

The requirements manager, project manager, lead client, delivery manager and senior business manager must sign off on the reviewed requirements in that particular order. It is not possible for all those approvals to take place simultaneously. Why? The project manager relies on the requirements manager's signature for approval, even though the project manager must carefully read the BRD before sign-off. The lead client also must read the BRD, but he or she does not necessarily look for the same details as the project manager. The delivery manager and the senior business manager may browse through the BRD, but they rarely read it thoroughly. They rely on the project manager and the lead client, respectively.

By signing off on the BRD, the approvers confirm the following:

- The requirements manager and project manager confirm that the requirements are complete, necessary, concise, design free, attainable, verifiable and traceable.
- The lead client confirms that the requirements are in compliance with the business intentions.
- The delivery manager confirms that the requirements are completely understood by the delivery team, feasible and ready for implementation.
- The senior business manager confirms that business requirements are aligned with the corporate business goals and corporate strategy.

Since the BRD approval process is sequential, it may take some time to collect all signatures. It is usually done in a scheduled face-to-face meeting or electronically via circulation of the BRD.

After the requirements manager and project manager sign the BRD, it will be forwarded to the next person in the chain with a request to approve within three to five days and an offer to answer all questions which may have arisen. It cannot be any less than three days, because everybody should have a reasonable amount of time to read the BRD. On the other hand, if there is no sign-off deadline, the BRD may *never* get signed off on. In the afternoon before the due date, the project manager calls the person currently holding the BRD and asks about his or her availability for a brief meeting on the following day to receive the signed-off BRD. If the sign-off is electronic, a reminder should alert the individual of a sign-off deadline the next day. This usually helps to get the job done.

Still, the sign-off process may take two weeks or longer. This time must be used to work on tasks started during the Conduct Requirements Review (R7) process, when the project flow branched to the Planning Frame. Those tasks include:

- Develop project plans for the Planning Frame and the preliminary project plan for the rest of the project. Those activities will take place at the same time as executing processes R7 and R8 in order to utilize resources and the waiting time for the BRD sign-off. Processes involved in building project plans are described in detail later in this book.
- Get the resource managers' commitment for the Planning Frame resources.
- Make appointments with resource managers and present them with resource requests.

All of this is required for presentation to the lead client for review and getting his or her recommendation to the senior business manager and the sponsor to authorize the project.

The BRD constitutes the requirements baseline. Scope changes are tracked and managed throughout the life of the project using the formal scope change request process, described in Chapter 10 (in Section III on the Planning Frame). The requirements traceability matrix must be updated when any of the requirements change. The scope change request must identify other requirements, other project elements or deliverables that may be affected by a change. The scope change process is established in the Planning Frame.

Upon completion of this process, a control point question (requirements approved?) is asked. If the BRD is not approved, the issue most probably is not the BRD itself, but other issues which manifest themselves in the BRD. For example, the reason may be lack of funds or the changing needs of the corporation. In this case, the project will be terminated. If the business wants to renew the project with a different scope, it must submit a new project request and benefit statement.

Conduct Kickoff Meeting (R9)

The kickoff meeting is the first activity of the approved project execution, beginning with the Planning Frame. The kickoff meeting should be scheduled for at least two hours. This meeting provides the opportunity to get the delivery team members and their business counterparts together, introduce and celebrate the new project and, most importantly, gain buy-in from participants and contributors. Usually, by the time the kickoff meeting occurs, not all project team members are assigned. However, by then the major players should have been identified and invited to attend.

The invitation to the meeting along with the agenda are sent at least two weeks in advance. It is customary to serve some refreshments (food and beverage) at a kickoff meeting, as it is actually a social event of sorts. In the first part of the meeting, the project manager should introduce everybody (or have attendees introduce themselves) and their roles. One idea is to have each person share the number of years of experience he or she has with similar projects. The project manager logs these numbers and then tells the team that they can rely on (for example) the century and a half of experience in the room. The project manager will then briefly describe the project goals and major deliverables. It is also necessary to describe the expected communications with the delivery team and client, such as status reports, and types and frequency of different progress meetings.

In the second part of the meeting, the participants are encouraged to get to know each other and enjoy the refreshments. Depending on the situation, an ice-breaker exercise can be facilitated by the project manager.

Review Planning Frame Milestones with Client (R10)

The Planning/High-Level Design Frame plan and the overall project plan now return to the Requirements Frame through entry point 8. The plan contains, among other documents, the schedule for the Planning Frame, along with the information about the cost and duration of the Planning Frame. Some of this information must be shared with the client. While the client does not have to know, for example, what it takes to run the quality assurance reviews or the duration of the statement of work development, the client definitely must be shown the project milestones, such as the project package completion date or the statement of work review date with the business. It is necessary to present a milestone for each approved requirement, along with milestones of other business-related activities, but not necessarily the procedural or technical activities.

As in other reviews, the list of participants must be determined and the related documentation must be sent to them in advance, along with the invitation. The review is conducted as a walk-through, reviewing milestones, the schedule and the budget for the Planning Frame, along with the preliminary budget and schedule through the end of the project. Notes should be taken. Any proposed changes will be reviewed with the delivery team members and the senior delivery manager in order to determine whether they are necessary and feasible to implement.

NOTE The general recommendation is not to show the client the detailed project schedule to avoid an attempt by the client to micromanage the project or a request to be informed of the completion of every scheduled task, which the client may not even understand. If you allow the client to see this level of project detail, the client may interfere with the project management processes, even asking to remove some quality- or risk-related tasks in order to save costs. Another reason is that you *cannot* show the client some of the project tasks, like internal project reviews and meetings or tasks related to containment of some negative elements related to the client in the project risk assessment. If the client is aware of those meetings, you cannot stop them from sending their representative. Instead of the detailed project schedule, provide the milestones important to *the client,* like client review dates, completion dates of client deliverables and the project completion milestones. However, as the one that pays for the project, the client may demand a complete schedule. In this case, the work breakdown structure should not directly show those confidential tasks or those which simply are not necessary for the client to see.

Update Planning Frame Plan (R11)

As the result of the review of the Planning Frame activities, some changes may be introduced to plans during their review and approval. Therefore, it is necessary to update plans and milestones and send changes to the review participants. Consent must be received before asking for project authorization by the project sponsor.

Request Project Authorization (R12)

During execution of this process, the project manager requests that the project sponsor or the lead client arrange for the project charter to be issued and obtains commitment to continue scheduled tasks before the kickoff meeting for reasons described earlier. The project manager should receive the project charter from the project sponsor or senior business manager. This is usually a short and high-level document which authorizes the project manager to continue project execution beyond the Requirements Frame. The project charter does not describe project details, but rather provides reference to the signed-off BRD.

If the project charter is not received, the project must not continue due to the exceptional risk involved, including the threat of the project manager simply not having proper authority to obtain resources for the project. After notifying all project stakeholders, the project will be terminated by directing the project flow to the Closing Frame via exit point 3.

Requirements Management Process Metrics

Requirements management process metrics are recorded while being tracked in the Construction Frame, like the process metrics of other project frames. Tracking implementation of the Requirements Frame involves *comparing the plan to actuals.* The weekly status throughout the requirements analysis is recorded and the following metrics are gathered:

- Plan versus actual date for completion of each requirement
- Plan versus actual date for completion of all requirements
- Plan versus actual date for completion of the BRD
- Plan versus actual date for the kickoff meeting

As mentioned earlier, the project flow returns to the Requirements Frame for the requirements analysis from all other project frames when a project scope change is initiated. Along with the comparison of the scope change plan versus actuals, important information about the *stability* of requirements is also obtained. If a requirement has too many changes throughout the project life cycle, this may mean that either the requirement was not gathered properly or that the client was not sure what he or she wanted from the project. Having those statistics will help to avoid similar issues in future projects.

When managing project scope changes throughout the project life cycle, the following metrics are recorded:

- Plan versus actual for completion of each project scope change
- Total cost expended for all changes by project frame
- Total effect of scope change requests on schedule by project frame
- Number of scope change requests issued by project frame
- Number of the scope change requests approved by project frame
- Number of changes to each requirement
- Total number of changes for the project

Requirements Frame Completion Criteria

The Requirements Frame is assumed to be complete when:

- The requirements management procedure is identified and documented in the project control book (PCB)

- All business requirements are documented in the BRD
- The BRD is filed in the PCB
- Evidence of requirements review is documented in the PCB
- BRD approval is documented in the PCB
- The requirements traceability matrix is documented in the PCB
- Metrics are documented in the PCB
- The project charter is received and documented in the PCB
- The kickoff meeting is conducted and the record of it is documented in the PCB

Even though the BRD is signed off on and the project charter is received, the Requirements Frame is not assumed to be complete until the above evidence is in place. It is the responsibility of the project manager and quality assurance analyst to ensure that this is the case.

Section III. Planning/High-Level Design Frame

Planning/High-Level Design Frame Process Flow

Introduction

The purpose of the project planning process is to develop plans for executing and controlling all project frames and processes within each frame. This chapter details how to break down project tasks, estimate them, package them and then, equally important, communicate the detailed project plan. The Planning Frame consists of the following major elements:

- Develop the high-level design and architecture, where appropriate
- Develop the communication management plan
- Develop the risk management plan
- Develop the quality management plan
- Develop the configuration management plan
- Develop the resource management plan
- Develop the outsourcing/offshore management plan
- Estimate project activities
- Initiate or update the work breakdown structure
- Develop or update the plan package
- Communicate the plan package
- Develop the statement of work
- Update and approve the project, frame and scope change budget

Table 5.1 Frames Planning Initiation Source

Planning of	Where the Planning Is Initiated
Requirements Frame	From the Requirements Frame soon after receiving initial project requirements.
Planning Frame	From the Requirements Frame at the time of the final requirements review. The plan details may be updated, if necessary, after the project Planning Frame approval, just before project authorization is issued.
Construction Frame	Internal loop from within the Planning Frame, after the Planning Frame plan is complete.
Closing Frame	Internal loop from within the Planning Frame, after approval of the Construction Frame plan and the beginning of its implementation and tracking.

The combination of the above elements is used to develop plans for all project activities that have a major impact on project cost, duration and quality. The more thorough the project plan, the more predictable the project cost and schedule.

All activities require planning. This includes planning of the Planning Frame, since the Planning Frame activities include plans for high-level design, plans for risk management, plans for quality management, etc. Planning of all project activities and the overall project is always done in the Planning Frame, but may be initiated in other frames. Where the planning of each frame is initiated is shown in Table 5.1.

NOTE The Construction Frame planning may start early, in the middle of the Planning Frame planning, but in order to complete the plan, it is necessary to be in the advanced stages of the Planning Frame planning. If most of the Planning Frame is not complete, there is not sufficient information available for planning of the detailed design and implementation, which are parts of the Construction Frame. Similarly, in order to complete the plan for the Closing Frame, it is necessary to be in the advanced stages of the Construction Frame implementation.

Roles and Responsibilities

The following roles and responsibilities will be required in the Planning/High-Level Design Frame:

- **Project manager**—Responsible for ensuring that all processes are adhered to and documented in the project control book. The project manager ensures that the project team is engaged in all stages of project planning. The project manager tracks the implementation of the Planning Frame activities and reviews the status on a weekly basis with the lead client and on a monthly basis with senior business and senior delivery management. In some environments, the project manager reports the status of the project to the senior business manager on a weekly rather than monthly basis. This is a matter of due diligence based on size and scope of the project.

- **Delivery team**—Also referred to as the project team, the assigned delivery team members will be responsible for producing documentation in the Planning Frame and outlining how the approved requirements will be implemented. They will produce the high-level design and participate in quality audits and reviews, as well as present individual weekly status reports to the project manager during the scheduled project status meeting.
- **Client**—This is the lead client and a single focal point for the business. The client should participate in all status review meetings scheduled by the project manager. The client is sometimes referred to as the business project manager or the business area lead. The client usually reports to the senior business manager.
- **Technical lead**—Leading technical specialist who is a member of the delivery team. The technical lead is in charge of producing the high-level design and architecture. He or she will assist the project manager in developing the work breakdown structure and gathering estimates from members of the delivery team.
- **Senior business manager**—The person whose signature is required on the statement of work. The senior business manager should review the monthly status report with the project manager. The senior business manager usually reports to the project sponsor.
- **Senior delivery manager**—The person who owns a delivery budget and whose signature is required on the statement of work.
- **Project sponsor**—Major stakeholder who is responsible for the business success of the project, specifically, ensuring that the business objectives for which the project has been undertaken are met. The project sponsor is the *owner* of the project.
- **Quality assurance analyst**—Responsible for running all quality management reviews and audits throughout the project life cycle in strict compliance with the quality management process. He or she is responsible for documenting reviews, rating them and for follow-ups when needed. The quality assurance analyst should be someone with quality assurance training who is not directly associated with the project. However, for smaller projects, the project manager may play the quality assurance analyst role.

Inputs/Outputs

The Planning Frame processes interact with all project frames via the entry and exit points, as shown in the Planning Frame process flow diagram in Figure 5.1:

- **Entry point 1**—
  - 1A—From the Requirements Frame for the requirements analysis plan
  - 1B—From the Requirements Frame for the project scope change plan
  - 1C—From the Requirements Frame for the Planning Frame plan
  - 1D—From the Requirements Frame for preliminary estimates of project costs
- **Entry point 2**—From the Requirements Frame for ballpark estimates of the project cost

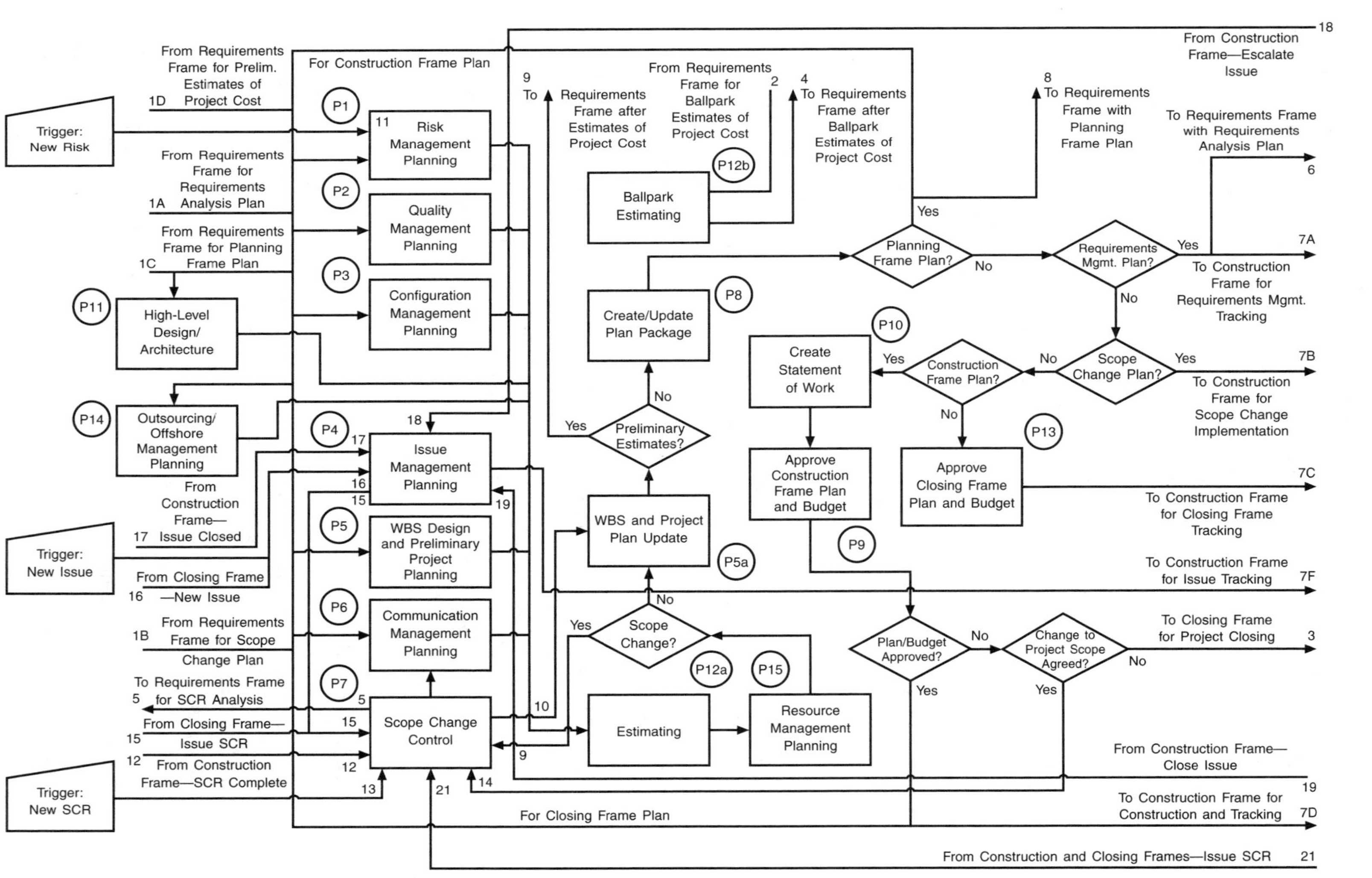

Figure 5.1 Planning Frame Process Flow Diagram

- **Entry point 12**—From the Construction Frame to provide notification that the scope change request (SCR) implementation is complete
- **Entry point 13**—From a new SCR trigger to open a new SCR
- **Entry point 16**—From a new issue trigger to open a new issue
- **Entry point 21**—From the Construction Frame to trigger a new SCR

The process flow is directed out of the Planning Frame through the following exit points:

- **Exit point 3**—To the Closing Frame for immediate project close
- **Exit point 4**—To the Requirements Frame with ballpark estimates of the project cost
- **Exit point 5**—To the Requirements Frame for SCR analysis
- **Exit point 6**—To the Requirements Frame with the requirements analysis plan
- **Exit point 7**—
  - 7A—To the Construction Frame for requirements management implementation and tracking
  - 7B—To the Construction Frame for scope change implementation and tracking
  - 7C—To the Construction Frame for Closing Frame implementation and tracking
  - 7D—To the Construction Frame for construction plan implementation and tracking
  - 7F—To the Construction Frame for issue resolution and tracking
- **Exit point 8**—To the Requirements Frame with the Planning Frame plan
- **Exit point 9**—To the Requirements Frame with preliminary estimates of the project cost

Planning Frame Process Flow

Requests to the Planning Frame

There are 11 different requests to the Planning Frame, which enter the frame via entry points 1, 2, 12, 13, 16 and 21 and through the internal loop from within the Planning Frame (see the Planning Frame process flow diagram in Figure 5.1):

1. Develop Requirements Frame plan (entry point 1A)
2. Develop plan for the Planning/High-Level Design Frame (entry point 1C)
3. Develop Construction/Tracking Frame plan (internal loop)
4. Develop Closing/Testing Frame plan (internal loop)
5. Develop scope change plan (entry point 1B)
6. Develop preliminary estimates of the project cost (entry point 1D)
7. Develop ballpark estimates of the project cost (entry point 2)
8. Receive notification that an SCR is complete (entry point 12)
9. Receive new SCR request (entry point 13)

10. Receive new issue request (entry point 16)
11. Trigger new SCR request from Construction Frame (entry point 21)

Requests 1 through 5 of the above requests are requests to produce project plans. Requests 6 through 11 are for other activities within the Planning Frame. The receive new SCR request (9), after initial processing, directs project flow to the Requirements Frame in order to produce SCR requirements analysis. When the analysis is complete, the request comes back to get a small or large scope change plan. A receive new issue request (10) needs the separate issue resolution plan made during the Issue Management Planning (P4) process. A trigger new SCR request initiates the trigger, which in turn issues a receive new SCR request (9).

There are nine processes at the beginning of the frame, which are called entry processes:

P1 Risk Management Planning
P2 Quality Management Planning
P3 Configuration Management Planning
P4 Issue Management Planning
P5 Work Breakdown Structure (WBS) Design and Preliminary Project Planning
P6 Communication Management Planning
P7 Scope Change Control
P11 High-Level Design/Architecture
P14 Outsourcing/Offshore Management Planning

Table 5.2 shows which planning processes are executed based on the request type:

- **must**—The specified entry process is always used.
- **may**—The specified entry process may or may not be used, depending on specific circumstances.
- **no**—The specified entry process is never used for that type of request.

The Issue Management Planning (P4) and Scope Change Control (P7) processes can be directly triggered only by a new issue and a new SCR. For example, even if a new issue comes up during requirements planning, the new issue trigger is initiated, which in turn generates a new issue request to the Issue Management Planning process.

Table 5.2 may be useful even if project management processes have already been established in the organization. Those processes must be universal enough to suit any project in the enterprise's portfolio. If those processes have not yet been established, the guidelines provided in the corresponding chapters of this book can be used to establish project management processes and document them in the project control book.

Once a process is planned, it must be executed at some point. Therefore, for each process planned, there must be a corresponding execution or management process. For example, after the quality activities are included in the approved plan and the work starts, *the quality must be periodically tracked and maintained.* The result, or outcome, of the

Table 5.2 Processes Utilized in Different Types of Requests

Requests	Entry Processes Used								
	Risk Management Planning (P1)	Quality Management Planning (P2)	Configuration Management Planning (P3)	Issue Management Planning (P4)	WBS Design and Preliminary Project Planning (P5)	Communication Management Planning (P6)	Scope Change Control (P7)	High-Level Design/Architecture (P11)	Outsourcing/Offshore Management Planning (P14)
Get requirements plan (entry 1A)	**must**	**must**	no	no	**must**	**must**	no	no	may
Get small scope change plan (entry 1B)	no	no	no	no	no	no	**must**	may	may
Get large scope change plan (entry 1B)	**must**	**must**	may	no	no	may	**must**	may	may
Get Planning Frame plan (entry 1C)	**must**	**must**	**must**	no	**must**	**must**	no	**must**	may
Get Construction Frame plan (internal loop)	**must**	**must**	**must**	no	**must**	**must**	no	no	may
Get Closing Frame plan (internal loop)	**must**	**must**	no	no	no	**must**	no	no	may
Get issue resolution plan (entry 16)	no	no	no	**must**	no	no	no	no	may
Get new SCR (entry 13)	no	no	no	no	no	no	**must**	no	may
Get preliminary estimates (entry 1D)	no	no	no	no	**must**	no	no	no	may
Get ballpark estimates (entry 2)	no	no	no	no	no	no	no	no	no

quality planning process is the quality plan. In order to maintain quality, the quality management process must be executed. It does not make sense to attempt process planning without a clear understanding of the process execution. It is not really possible to plan risk management if we do not know how to manage risks.

The following section presents details of the risk, quality, communication and other management processes. Some of these processes will be executed in the Construction Frame of the project, and all of them will be tracked in the Construction Frame.

Process Overview

The overall Planning Frame process flow diagram is shown in Figure 5.1. During multiple runs of the Planning Frame, the following processes are executed:

1. **Risk Management Planning (P1)**—Formally, the purpose of risk management is to identify and document risks (which include negative risks or *threats* and positive risks or *opportunities*), to control the impact of risks and to minimize the effects of threats and maximize the effects of opportunities on the project. The Risk Management Planning process produces the risk containment plan, which may affect project estimates. However, only negative risks will be considered in this book for reasons described later.
2. **Quality Management Planning (P2)**—The Quality Management Planning process will establish quality management plans and schedule quality assurance audits and quality control reviews. Quality management will control quality throughout the project life, from inception to closing. Without this process, there would be no way to determine whether or not the project implementation is successful. Quality management should also be focused on the quality of the project process itself. By reading this book and taking our advice about project workflow, you are actually already improving this aspect of quality management!
3. **Configuration Management Planning (P3)**—The purpose of this process is the control of project components at any given time, preventing unauthorized changes to the completed deliverables. The Configuration Management Planning process produces the schedule of configuration management reviews. Schedules are part of the project plan package, which is stored in the project control book.
4. **Issue Management Planning (P4)**—The purpose of issue management is to identify and manage issues that come up during all project frames, to identify processes to resolve issues and minimize their impact on the project. An issue differs from a risk in that it is a 100% certain threat to the project, and it should be managed somewhat differently because of that element of certainty. The issue management plan identifies resources responsible for each issue resolution task, resources for escalation when needed and the target dates for the issue resolution. The Issue Management Planning process is triggered every time an issue comes up. In some cases, when an issue cannot be resolved without a scope change, it triggers a new SCR.
5. **WBS Design and Preliminary Project Planning (P5 and P5a)**—The purpose of a WBS as it is used in this book is to enable the establishment of plans for

managing the project or frame. The WBS will contain the project information (e.g., milestones, deliverables, dependencies, risks, tasks, etc.) and identify resource requirements and training plans. The WBS incorporates inputs from processes P1 through P6, P11 and P14. Changes to the project scope during other project frames will bring the project flow back to this process.

6. **Communication Management Planning (P6)**—Communication management is required to identify stakeholders and develop project reporting and the reporting templates for different types of project communication. Communication planning produces the scheduled communications to stakeholders, as well as the choice of medium for the various project communications.
7. **Scope Change Control (P7)**—The project Scope Change Control process establishes rules for implementing scope changes, while avoiding scope leak and scope creep. Scope leak occurs when any planned work on the approved baselines is omitted or deferred to a future time frame without proper change authorization. Scope creep occurs when any unplanned work on the approved baselines is performed without proper change authorization. In other words, scope leak occurs when some project scope elements are missing in the end product, whereas scope creep occurs when additional undocumented scope elements present in the end product or the delivered project scope differ from the documented one. Scope change planning produces the implementation schedule for each formal SCR. SCRs must be received before planning scope change.
8. **Create/Update Project Plan Package (P8)**—The project plan package is a set of the various plans, schedules, resource assignments, estimates, etc. required for project or frame budget approval.
9. **Approve Construction Frame Plan and Budget (P9)**—This step is executed at the end of the Planning Frame just before the Construction Frame starts. The final frame budget must have a definitive accuracy of –5% to +10%.
10. **Create Statement of Work (P10)**—The purpose of this process is to describe the steps necessary to create the statement of work, which is the most important legal document as it lays the foundation for the project. As opposed to the charter, this is a narrative description of the work to be done and schedule of payments.
11. **High-Level Design/Architecture (P11)**—This process will start execution at the same time with all the planning elements (P1 through P6) and must be finished before the plan package is produced in process P8. The flow for process P11 is purely technical in nature and will be entirely different for different industries.
12. **Estimating (P12a) and Ballpark Estimating (P12b)**—The purpose of estimating is to produce size, effort, cost and critical element estimates for a project or a frame throughout its life cycle. The Estimating (P12a and P12b) process describes necessary activities and methods required to produce estimates. Process P12b produces only ballpark estimates, while P12a produces either preliminary estimates for the entire project or definitive estimates for the project frame being planned.
13. **Approve Closing Frame Plan and Budget (P13)**—This process is executed at the end of the Construction/Tracking Frame, just before the Closing/Testing

Frame starts. The final budget determined here has an estimated definitive accuracy.

14. **Outsourcing/Offshore Management Planning (P14)**—The purpose of outsourcing management is the selection of qualified suppliers (outside of the project team) for implementation of project components, as well as managing the relationship with them for quality deliverables and seamless integration with other components of the project.
15. **Resource Management Planning (P15)**—The purpose of resource management is to manage resources (specifically people) on the project.

The issue management and scope change plans are both dynamic plans, which are built only when a new issue comes up or a scope change is requested in any project frame. Both must be documented prior to developing plans.

There are seven different process flow paths in the Planning Frame, one for each type of planning request:

1. Requirements Frame planning path (entry point 1A)
2. Scope change planning path (entry points 1B, 13, 14)
3. Planning Frame planning path (entry point 1C)
4. Construction Frame planning path (internal loop)
5. Closing Frame planning path (internal loop)
6. Preliminary or detailed project, frame or scope change estimates path (entry point 1D)
7. Ballpark project estimates path (entry point 2)

In addition, when a request is submitted for the issue resolution plan, the process flow follows the path described in the Issue Management Planning (P4) process. Each of the planning path flows will be reviewed separately.

Requirements Frame Planning Path

The Requirements Frame planning path is shown in Figure 5.1A (in bold). The process starts when the request from the Requirements Frame is sent out via exit point 1A to the Planning Frame early in the Requirements Frame, asking for the Requirements Frame plan to be produced. The process flow enters the Planning Frame via entry point 1A. Based on Table 5.2, the process involves the following entry processes:

- **Risk Management Planning (P1)**—The first project risk assessment must be made at the beginning of the requirements analysis and a second time before approval of requirements. The risk management plan will identify steps required to contain and remediate risks identified.
- **Quality Management Planning (P2)**—The quality management plan will identify quality audits and reviews to ensure the quality of requirements.
- **WBS Design and Preliminary Project Planning (P5)**—This process will provide the list of major activities needed to implement the Requirements Frame. The

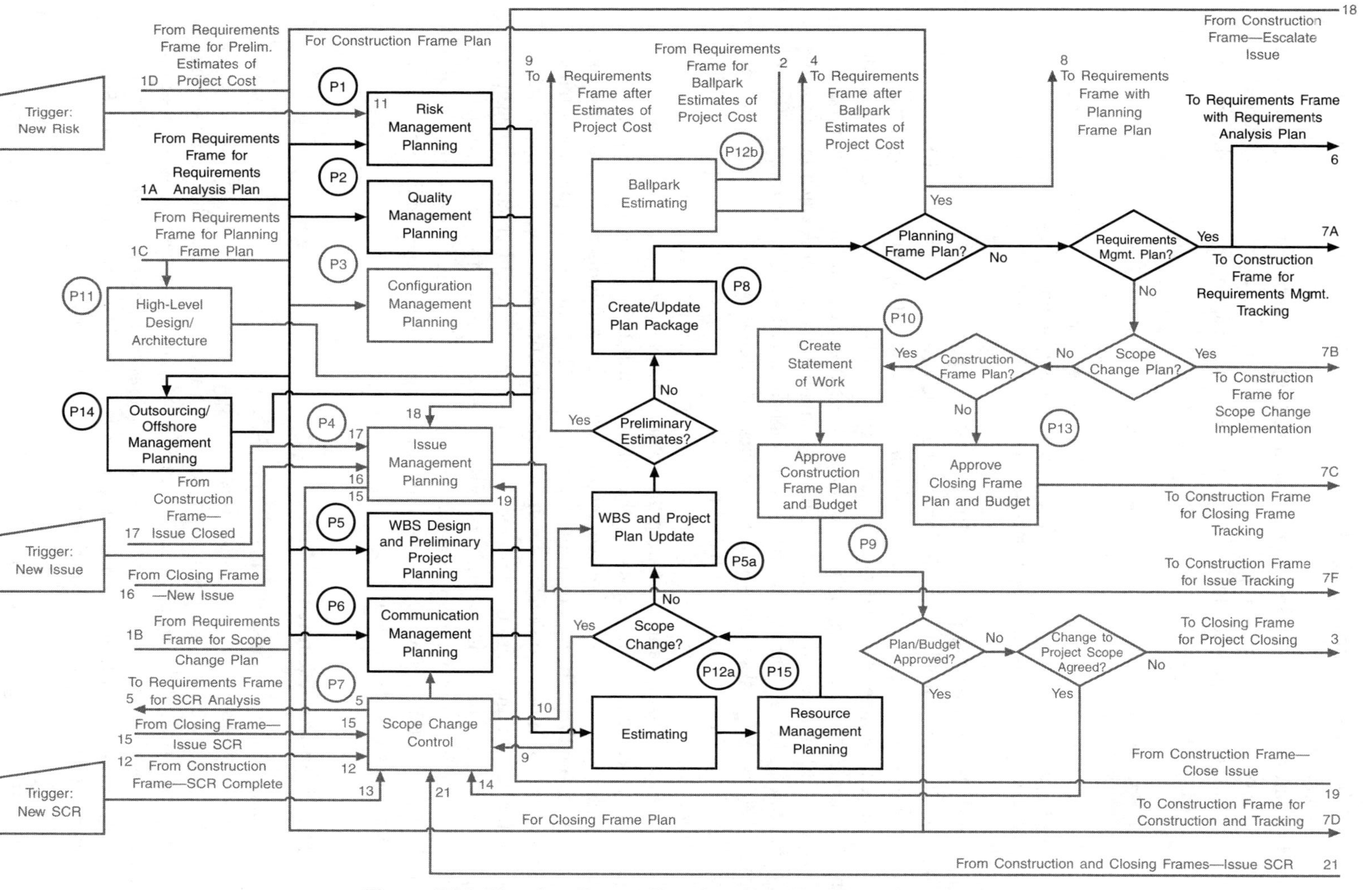

Figure 5.1A Planning Frame: Requirements Frame Planning Path

WBS will be updated later in the frame to produce a detailed plan for the requirements analysis.

- **Communication Management Planning (P6)**—The communication planning process will identify all project sponsors and describe interaction and reporting requirements between the delivery team and major sponsors during the requirements analysis.

NOTE Processes not in bold in Figures 5.1A to G are not relevant in the Requirements Frame planning flow. Also, scope change and issue management are event-driven processes, activated by the corresponding triggers. When triggered during requirements planning, they present separate parallel flow paths, which are not a part of this route.

When the execution of processes P1, P2, P5 and P6 is complete, the essential information will become available to produce new detailed estimates in process P12a and for resource planning in process P15. Since in this particular situation we are building the Requirements Frame plan and not the scope change plan, the answer to the decision point question (scope change?) will be no, and the project plan for the Requirements Frame will be updated in process P5a. It will contain a list of the detailed tasks for the Requirements Frame. During this process, all high-level tasks will be decomposed to the level of elementary tasks, with task dependencies indicated in the plan. Also, all resources will be planned and assigned by name to each task during execution of process P15.

From there, the process flows to a decision point question (preliminary estimates?). Since our goal at this point is building the Requirements Frame plan, rather than doing preliminary estimates, the answer to the question will be no, and the Requirements Frame plan package will be created in process P8 by combining all plans into one package. From there, the next decision point question (Planning Frame plan?) is reached. Since we are planning requirements management and not the Planning Frame plan, the answer is no. The answer to the next decision point question (requirements management plan?) is yes, and the process flow returns to the Requirements Frame via exit point 6, with the Requirements Frame plan available. At the same time, the process flow also goes to the Construction Frame for implementation and tracking of the requirements management activities via exit point 7A.

Scope Change Planning Path

The scope change planning path is shown in Figure 5.1B (in bold). A new SCR may be generated any time a new request arrives for scope change. In this case, the external trigger initiates a new SCR. The request to open a new SCR enters the Scope Change Control (P7) process via entry point 13. An SCR may also be generated as the result of certain conditions in the project process flow, such as budget or schedule issues, which cannot be resolved without reducing the project scope. The request enters the Scope Change Control (P7) process for the initial SCR review from:

- The Planning Frame via entry point 14, when the scope change is agreed to in order to reduce the budget or shorten the schedule

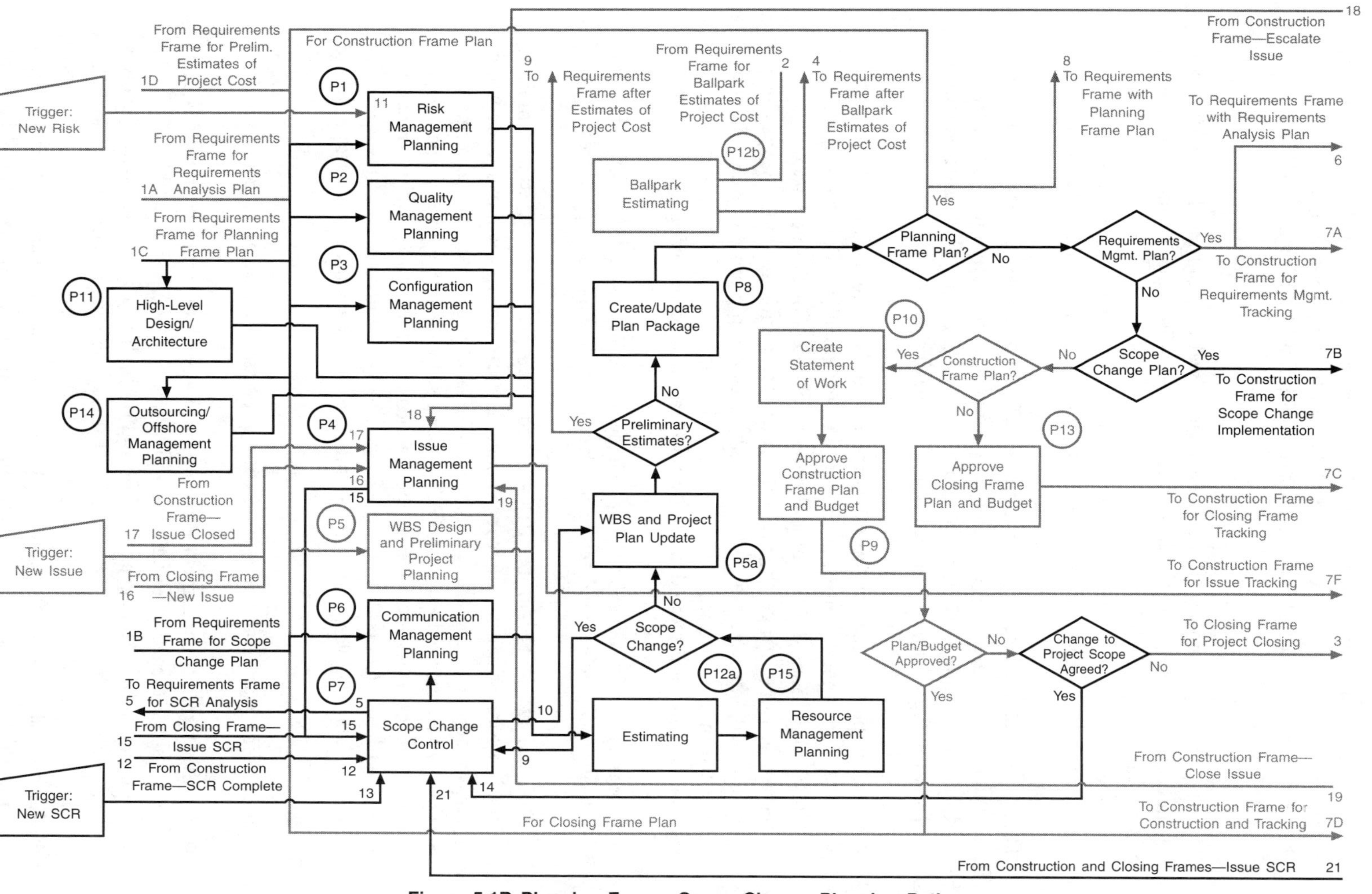

Figure 5.1B Planning Frame: Scope Change Planning Path

- The Issue Management Planning (P4) process via entry point 15, when the new request cannot be resolved without changes to the project scope
- The Construction or Closing Frame via entry point 21 during the troubled project assessment

The Scope Change Control (P7) process is described in detail in Chapter 10. If the SCR is not rejected there, then the flow goes to the Requirements Frame via exit point 5 to begin the SCR analysis.

After the SCR analysis is complete, the process flow comes back from the Requirements Frame to the Planning Frame via entry point 1B in order to produce the project scope change plan.

There are small and large scope changes, which do not always correspond to the size of actual changes and thus require different approaches. Sometimes it is not a straightforward task to determine exactly whether a scope change should be considered large or small. The overall guidelines to indicate *small* scope change are:

- Small scope change takes up to eight hours of labor to accomplish.
- Small scope change must not present any additional risks to the project.
- Small scope change does not add new stakeholders or business users.
- For small scope change, the SCR impact analysis performed in the Scope Change Control (P7) process indicates that nothing else in the project or outside of it is affected by the change.
- No *new issues* are created as the result of a small scope change.

If there is any doubt, consider a scope change to be large as a default. According to Table 5.2, *large* scope change planning involves the following entry processes:

- **Risk Management Planning (P1)**—The complete project risk assessment must be made before estimating the scope change. The risk management plan will identify steps required to contain and remediate risks of the scope change, if any.
- **Quality Management Planning (P2)**—A large scope change requires a separate quality assurance review, which must be planned. The project plan will be updated.
- **Scope Change Control (P7)**—The Scope Change Control process manages interaction details between the delivery and client teams when a project scope change is required.

Also, in some cases, three additional entry processes may be executed:

- **Configuration Management Planning (P3)**—If a large scope change adds extra deliverables, then the configuration plan must be updated as well.
- **Communication Management Planning (P6)**—If a large scope change involves additional business units or additional users, the existing communication plan must be updated to reflect the additional stakeholders.
- **High-Level Design/Architecture (P11)**—If a large scope change involves changes in the high-level design or architecture, then the required changes will be made using process P11.

- **Outsourcing/Offshore Management Planning (P14)**—If a scope change is related to or affects subcontractors, this process will be used for the scope change evaluation.

In the case of a small scope change, the Scope Change Control (P7) process is the only entry process executed.

When entry processes P1, P2, P3 and possibly P6, P11 and P14 are executed, the change request will be estimated in the Estimating (P12a) process and resources planned in the Resource Management Planning (P15) process. The process flow then reaches a decision point question (scope change?). Since the answer is yes, the process flow returns to the Scope Change Control (P7) process for SCR approval, as described in Chapter 10, and then sends the process flow to the WBS and Project Plan Update (P5a) process. From there, it follows the same path as in the Requirements Frame plan flow until it reaches a decision point question (requirements management plan?). This time the answer is no. Since we are in the process of scope change planning, the answer at the next decision point question (scope change plan?) is yes, and the process flow is directed to the Construction Frame for the scope change implementation and tracking via exit point 7B.

When the scope change implementation is complete, the Construction Frame sends notification of that to the Scope Change Control (P7) process via entry point 12 and the SCR is closed. The process flow returns to the beginning of the loop, waiting for the next change request.

The process flow for a small scope change is the same as for a large scope change, except that *no entry processes are involved*, except for the Scope Change Control process.

Planning Frame Planning Path

The Planning Frame planning path is displayed in Figure 5.1C. The request for the planning/high-level design path is triggered during the Conduct Requirements Review (R7) process in the Requirements Frame in order to obtain the following two plans just in time before project authorization is requested:

- Detailed Planning/High-Level Design Frame plan
- High-level project plan to the end of the project

The request, which enters via entry point 1C, activates the following entry processes:

- **Risk Management Planning (P1)**—Risk management plans are built and/or updated many times throughout the project. At the very least, they are refreshed at the beginning of each frame planning. It really is critical to remember that risk must be considered "live" and ongoing and not a one-time analysis at the start of the project.
- **Quality Management Planning (P2)**—Quality assurance reviews will take place at least twice during this frame, as described in Chapter 14.
- **Configuration Management Planning (P3)**—Configuration management reviews are usually done at least quarterly in accordance with the description in Chapter 8.

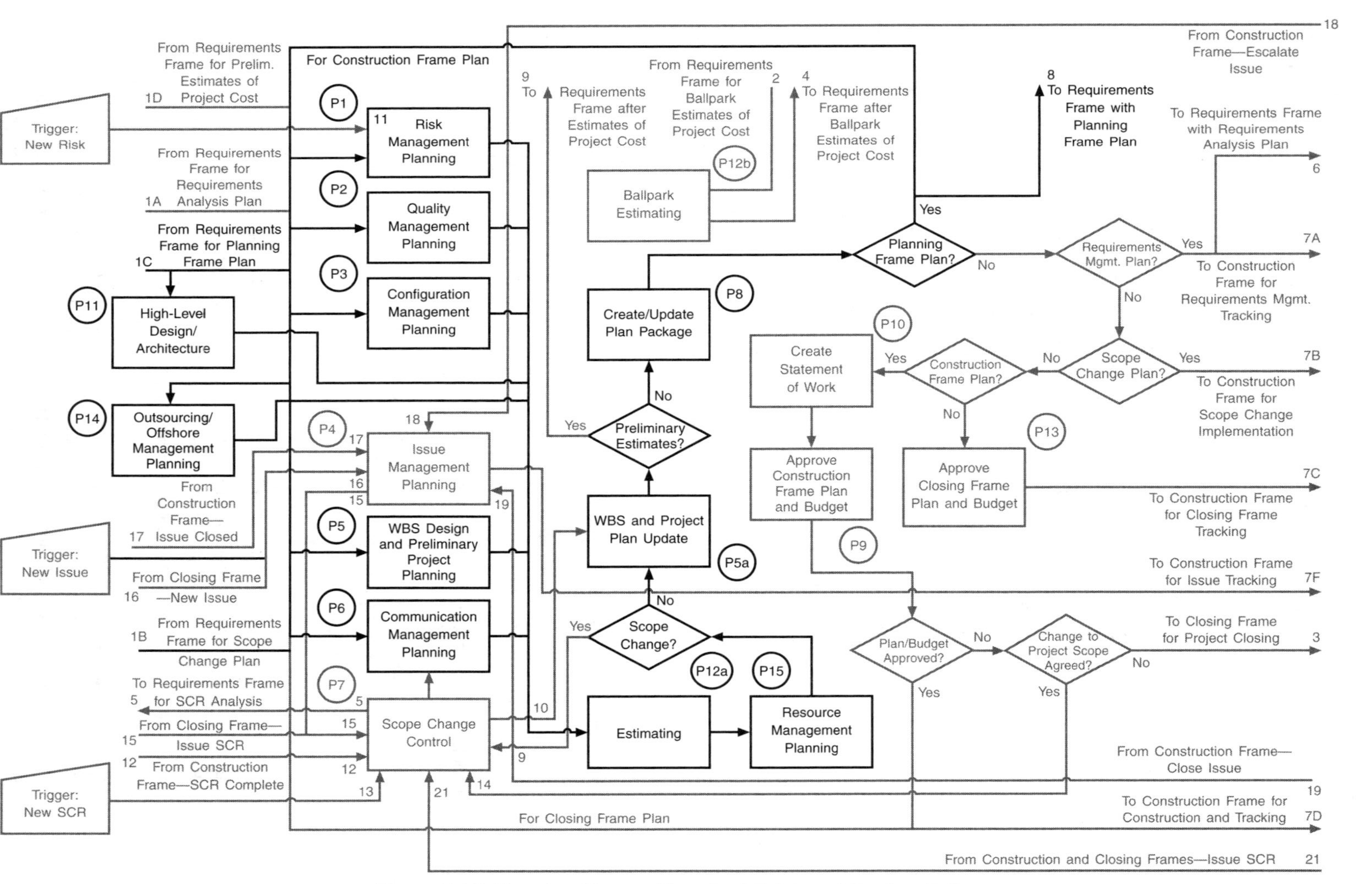

Figure 5.1C Planning Frame: Planning/High-Level Design Path

- **WBS Design and Preliminary Project Planning (P5)**—This process will provide the list of major activities needed to implement the project. The WBS will be updated later in process P5a to produce the detailed frame and the overall project plan.
- **Communication Management Planning (P6)**—The communication planning process will identify all project sponsors and describe interaction and reporting requirements between the delivery team and major sponsors during the Planning Frame.
- **High-Level Design/Architecture (P11)**—This is the technical process that guides the delivery team. The process depends on the industry and will differ significantly for different types of projects. Projects in process improvement and some other areas will have neither high-level design nor architecture.
- **Outsourcing/Offshore Management Planning (P14)**—Plans made by a subcontractor will be incorporated in the plan package in process P8.

When the execution of entry processes P1, P2, P3, P5, P6, P11 and P14 is complete, the process flow enters the Estimating (P12a) process. From this moment on, the process flow follows the same path as in the Requirements Frame planning flow up to a decision point question (planning frame plan?). Since this is indeed what is going on at the moment, the answer is yes, and the flow is sent back to the Requirements Frame with the completed Planning/High-Level Design Frame plan and the overall project plan. At the same time, the flow loops back to the beginning of the Planning/High-Level Design Frame to initiate the Construction/Tracking Frame planning.

Construction Frame Planning Path

When the Planning/High-Level Design Frame planning is complete and its implementation and tracking start in the Construction Frame, the Construction Frame planning is initiated when the process flow loops back to the beginning of the Planning Frame. The Construction Frame planning path is shown in Figure 5.1D. The Construction Frame planning request utilizes the following entry processes:

- **Risk Management Planning (P1)**—Risk assessments will be planned for the Construction Frame. They take place at least once at the beginning of the Construction Frame planning. If new risks are identified, a risk management plan must be built.
- **Quality Management Planning (P2)**—Quality audits and reviews will be planned for implementation, as described in Chapter 14.
- **Configuration Management Planning (P3)**—Since the configuration management reviews are usually done at least quarterly, it may or may not be necessary to plan a review in the Construction Frame.
- **WBS Design and Preliminary Project Planning (P5)**—The Construction Frame planning process will provide a list of activities needed to implement the Construction Frame. The previously created preliminary project plan will be updated to produce the detailed plan.

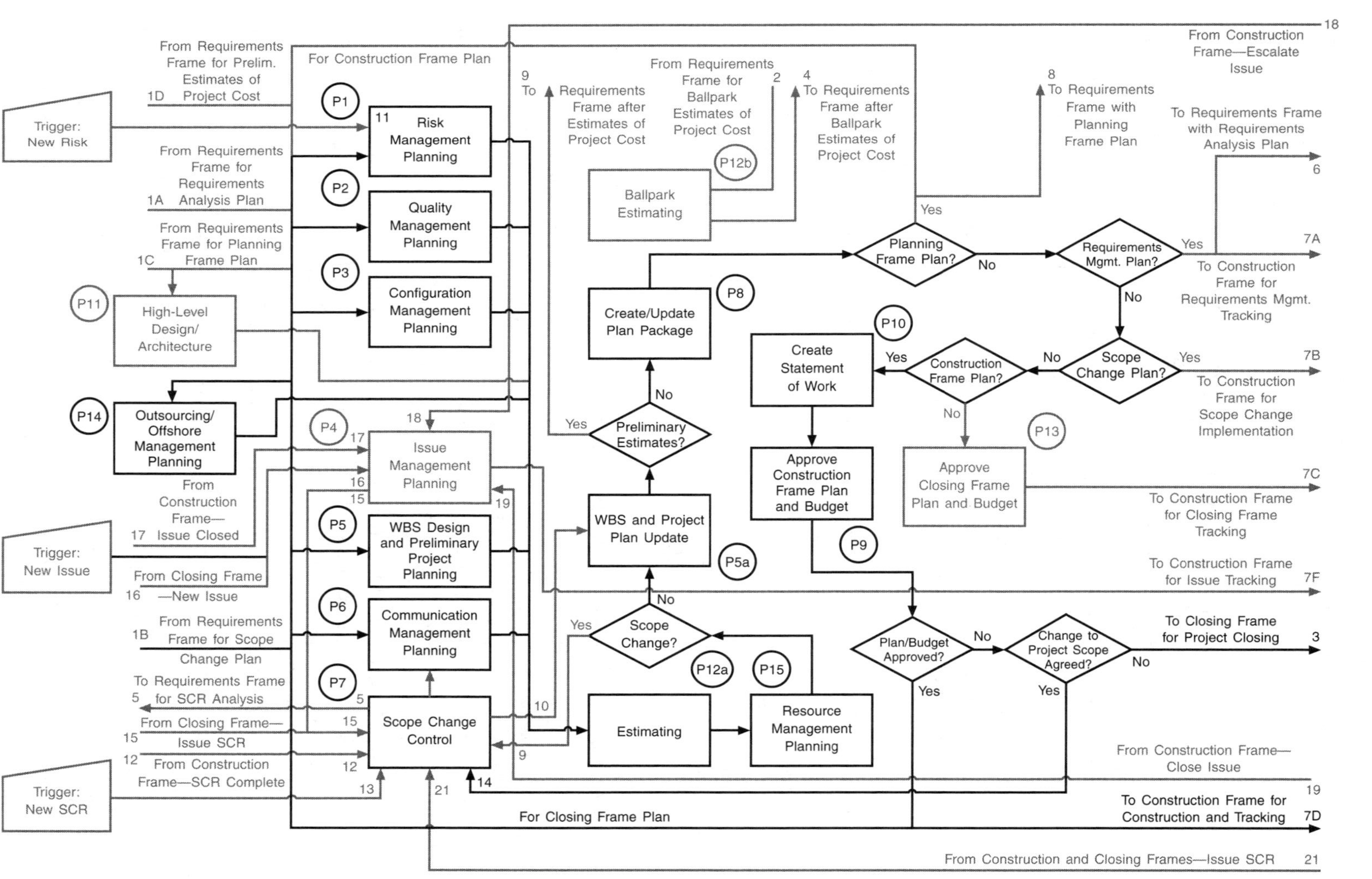

Figure 5.1D Planning Frame: Construction Frame Planning Path

- **Communication Management Planning (P6)**—The Communication Management Planning process will be updated to identify additional project stakeholders, if any, and describe interaction and reporting requirements between the delivery team and those stakeholders during the Construction Frame.
- **Outsourcing/Offshore Management Planning (P14)**—Detailed plans made by subcontractors will be incorporated in the plan package in process P8.

From the moment the execution of entry processes P1, P2, P3, P5, P6, P7 and P11 is complete, the process follows the same path as the Planning Frame path until the process flow gets to a decision point question (Planning Frame plan?). Since the Construction Frame is being planned at this instance, the answer is no. The next two decision point questions (requirements management plan? and scope change plan?) will both be answered no. When the flow gets to the next decision point question (Construction Frame plan?), the answer is yes, and the process flow goes to the Create Statement of Work (P10) process for development of the statement of work for the project, which is a legal document that guides the relationship with the client. The statement of work is developed in the Planning Frame, when planning for the Construction Frame is largely completed and the costs and schedule through the end of the project can be reasonably estimated. The statement of work provides a narrative project description, implementation milestones, price, assumptions, terms of payments, etc. Details of statement of work development are provided in Chapter 13.

When the statement of work is completed, the process flow goes to the Approve Construction Frame Plan and Budget (P9) process for the sponsor's approval.

The next decision point question is the plan (plan/budget approved?). If the plan is not approved, the reason must be analyzed. If the cost of project implementation is too high for the client or if the schedule is unacceptable, the client may agree to consider reducing the scope of the project in order to fit the acceptable budget or schedule. Any such move will require issuing a commensurate SCR. If the client agrees to the reduced project scope, then the answer to the decision point question (change to project scope agreed?) is yes. The process flow goes to the new SCR trigger to start the SCR process. However, if the scope change is not agreed to by the sponsor, the process flow is directed to the Closing Frame for project termination via exit point 3.

If the plan and budget are approved, the process flow goes to the Construction Frame for implementation and tracking via exit point 7D. At the same time, the process flow loops back to the beginning of the Planning Frame for planning of the Closing Frame.

Closing Frame Planning Path

The process is initiated by the process flow looping back to the beginning of the Planning Frame when the Construction Frame plan is approved. The Closing Frame planning path is shown in Figure 5.1E.

The request utilizes the following entry processes:

- **Risk Management Planning (P1)**—The risk management plan will be built at least once at the beginning of the Closing Frame planning.

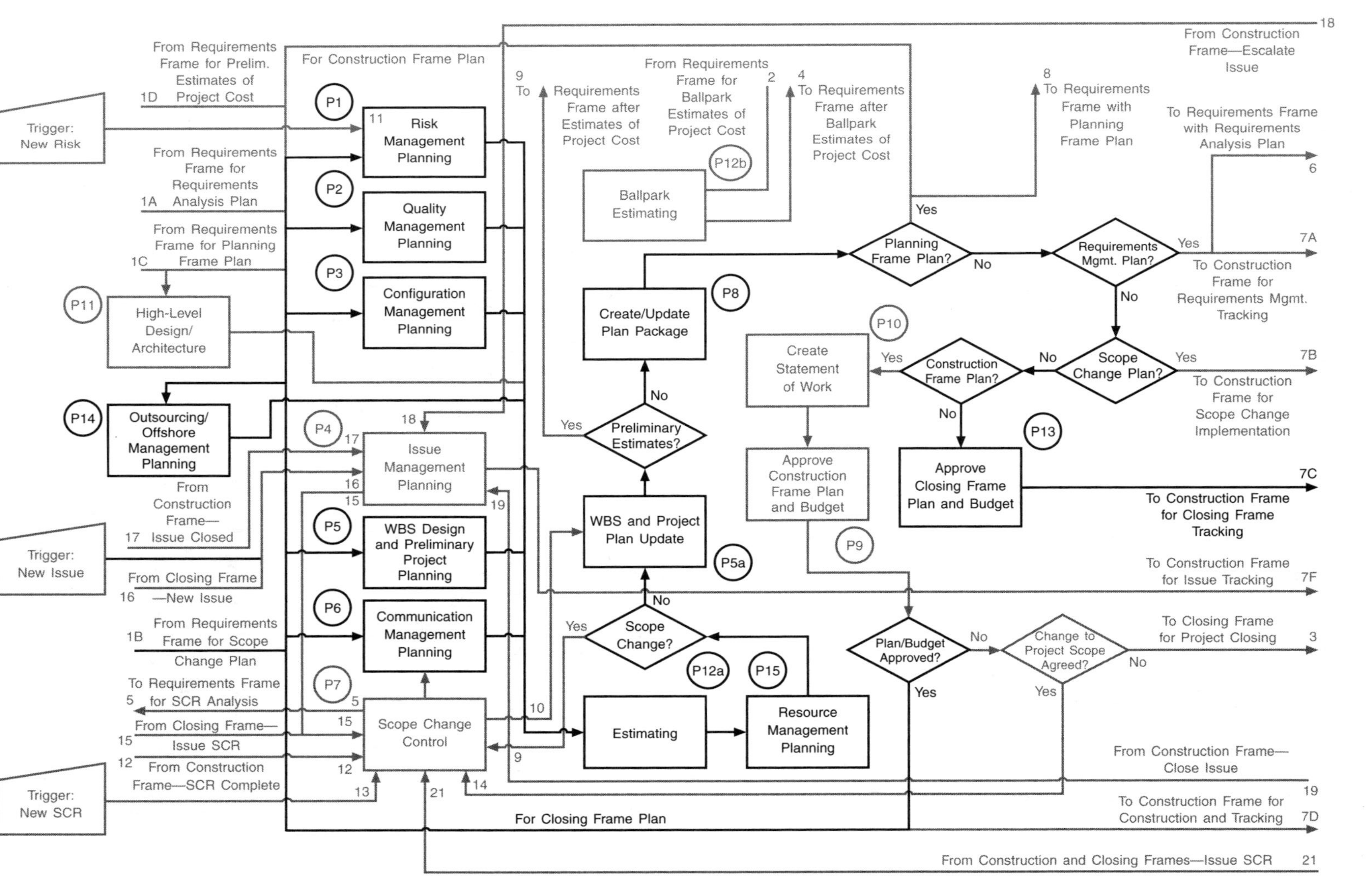

Figure 5.1E Planning Frame: Closing Frame Planning Path

- **Quality Management Planning (P2)**—Quality audits and reviews will be performed at least twice, as described in Chapter 14.
- **Configuration Management Planning (P3)**—Since the configuration management reviews are usually done at least quarterly, it may or may not be necessary to execute process P3.
- **WBS Design and Preliminary Project Planning (P5)**—This process will provide a list of activities needed to implement the Closing Frame. The preliminary WBS will be updated in process P5a to produce the detailed plan. While planning for the Closing Frame, there will no longer be any preliminary project planning, since that is the last frame of the project.
- **Communication Management Planning (P6)**—The communication planning process will be updated to identify additional project stakeholders and describe interaction and reporting requirements between the delivery team and those stakeholders during the Closing Frame.
- **Outsourcing/Offshore Management Planning (P14)**—Detailed Closing Frame plans made by a subcontractor will be incorporated in the plan package in process P8.

The process flow follows the same path as in the scope change plan until the flow comes to a decision point question (scope change plan?). The answer is no. At the next decision point question (Construction Frame plan?), since at this point the Closing Frame is being planned, the answer to the question will be no, and the process flow goes to the Approve Closing Frame Plan and Budget (P13) process for Closing Frame approval.

The Closing Frame is the last frame and is relatively small, so that the option of not approving the Closing Frame plan when all the other plans are approved cannot practically happen. The possible issue of contention in this frame is a user acceptance test, which may be resolved in the issue management process, as described in Chapter 8. Once the Closing Frame plan and budget are approved, the process flow goes to the Construction Frame for tracking of the Closing Frame and to the Closing Frame for the user acceptance tests and project closing.

Preliminary Estimates Path

The preliminary estimates path utilizes the following entry processes:

- **Risk Management Planning (P1)**—The risk management plan will be developed here to identify the risk margin of the preliminary estimates.
- **WBS Design and Preliminary Project Planning (P5)**—This process will provide a high-level list of activities for each project requirement. The list will include all project frames.

The preliminary estimates planning path is shown in Figure 5.1F. The process is triggered from the Requirements Frame when the business requirements document is produced in process R6 and enters the Planning Frame via entry point 1D. High-level estimates, which are based on the preliminary project plan, will be produced in process

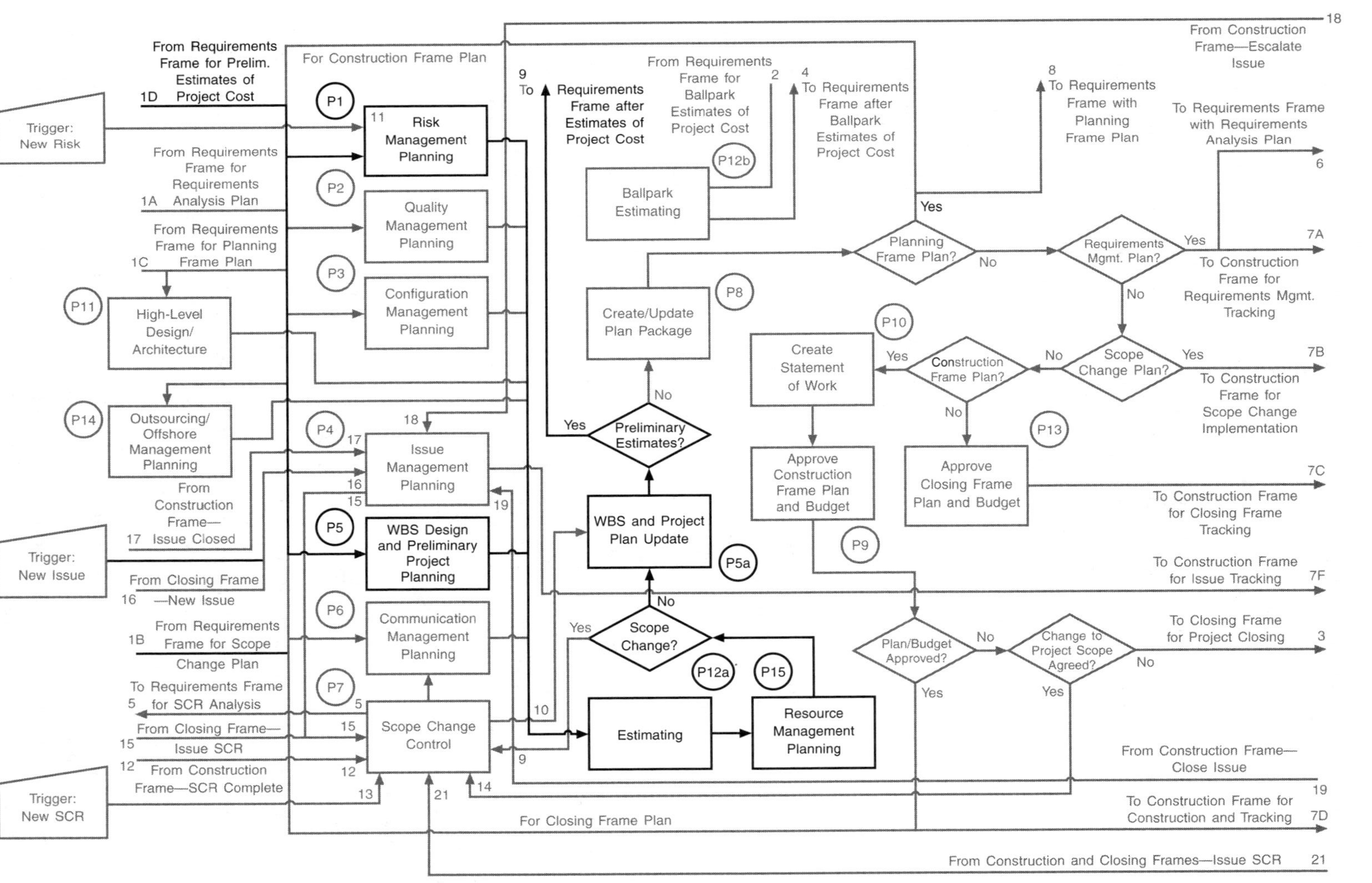

Figure 5.1F Planning Frame: Preliminary Estimates Planning Path

P12a. Later in the process, after detailed planning of the Planning Frame, when the Construction Frame and then the Closing Frame are complete, the Estimating (P12a) process will produce detailed estimates of the corresponding frames, one at a time. Next, the process flow reaches a decision point question (scope change plan?), and the answer is no. Following this, the WBS will be updated in process P5a on a very high level without detailed tasks, but with the high-level task dependencies identified and generic resources with no specific names assigned. While the Estimating (P12a) process will provide the effort for each activity, the task dependencies and resource assignments are required to calculate the overall approximate duration of the project, which will depend on the assumption of resource availability. The process flow reaches a decision point question (preliminary estimates?). The answer to this question is yes, and the process flow is returned to the Requirements Frame via exit point 9.

Ballpark Estimates Path

This process is initiated at the beginning of the Requirements Frame after receiving the initial project request. The ballpark estimating request enters the Planning Frame via entry point 2. The only process involved in ballpark estimates is the Ballpark Estimating (P12b) process. Upon completion of the requested estimates, the process flow returns to the Requirements Frame via exit point 4. The ballpark estimates planning path is shown in Figure 5.1G.

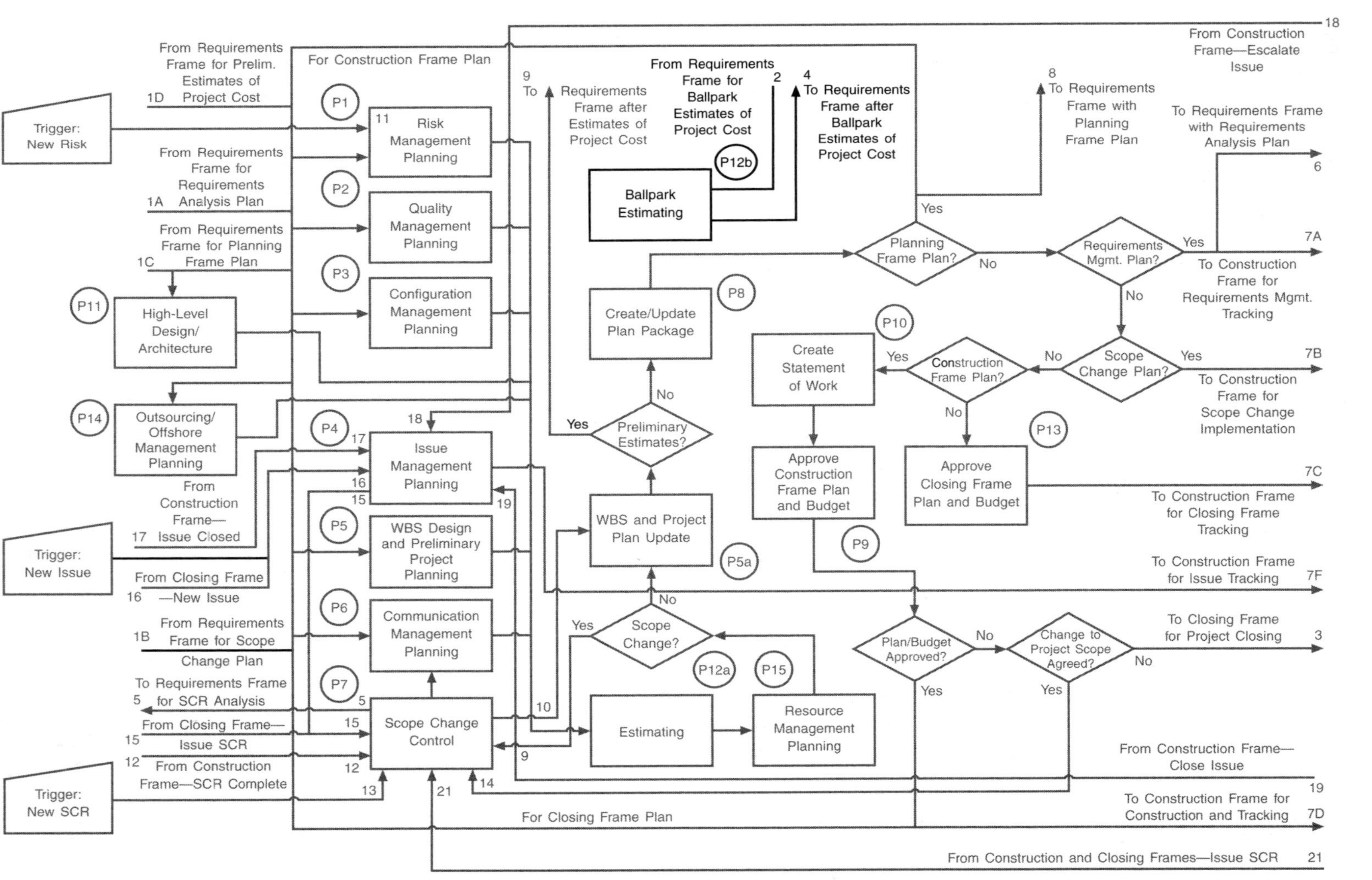

Figure 5.1G Planning Frame: Ballpark Estimates Planning Path

6

Risk Management (P1)

Purpose

The formal purpose of the risk management process is to ensure that all potential project risks (both threats and opportunities) are identified, their impact analyzed and risk management plans developed and implemented in order to eliminate or minimize the effects of threats and maximize the effects of opportunities on the project throughout its life cycle.

Since threats by far outweigh opportunities in frequency and effect on projects and since the vast majority of your work as a project manager in the real world will be to deal with threats, the following description of the Risk Management Planning (P1) process will deal only with negative risks, which for the purposes of this book can be read as "threats."

Risk Management Planning (P1) provides a risk containment or response plan, which may change project estimates. Project risk is an uncertain event, which, if it occurs, may have a negative impact on project deliverables in terms of the project cost, schedule and quality.

Risk management consists of three elements:

1. **Risk assessment**—The process of identifying and analyzing risks.
2. **Risk response planning**—The main purpose is to avoid or minimize harmful effects of threats on project cost, schedule and quality. This involves planning responses to high-scoring risks (those with high impact and probability of occurrence) or reducing their probability and/or impact. It also includes responses to low- and medium-scoring threats.
3. **Risk monitoring**—The process of monitoring risk occurrence, adjusting the project plan for the response to each of the risks identified and tracking the result of the

risk response. When a risk occurs, it is initially treated as an issue. As described in the discussion on issue management in Chapter 8, some issues require generating a project scope change request and treating the risk containment plans as scope changes in accordance with the Scope Change Control (P7) process (see Chapter 10). Risk monitoring is further described in Chapter 15 on the Construction Frame.

Risk Assessment and Risk Planning Process Flow

The process of risk assessment and planning is repeatedly executed many times during the project life cycle. It is executed once before the beginning of each project frame and also when significant project events happen during project execution, such as major issues, large scope change requests or unsatisfactory project performance. By implementing the process, all significant threats should be eliminated or reduced to the level of low risks, because the project cannot start or continue when a high threat level challenges not only the project but perhaps the entire organization. The process described here is executed in accordance with the risk assessment schedule, which is a part of the overall project schedule, as well as whenever an additional risk is identified during project execution.

The process flow is shown in Figure 6.1. The process consists of the following elements:

1. Identify Potential Risks (P1-1)
2. Determine Probability of Occurrence (P1-2)
3. Determine Maximum Loss Value of Each Risk (P1-3)
4. Calculate Expected Monetary Value (EMV) of Each Risk (P1-4)
5. Develop Risk Response Plan for Each Risk (P1-5)
6. Balance EMV of Acceptable Project Risks (P1-6)
7. Calculate Total Risk Costs (P1-7)
8. Determine Severity of Each Risk (P1-8)
9. Calculate Each Risk Rating (P1-9)
10. Balance Ratings of Acceptable Project Risks (P1-10)

Risk effect may be calculated by one of the following two methods:

1. Calculate project risk rating and convert it into a percentage of the project cost (processes P1-1, P1-2, P1-5, P1-7, P1-8, P1-9 and P1-10)
2. Calculate the EMV of risks in the project (processes P1-1, P1-2, P1-3, P1-4, P1-5, P1-6 and P1-7)

Processes P1-1, P1-2, P1-6 and P1-7 are used in both methods.

The Identify Potential Risks (P1-1) process starts when a risk assessment is triggered by one of the following:

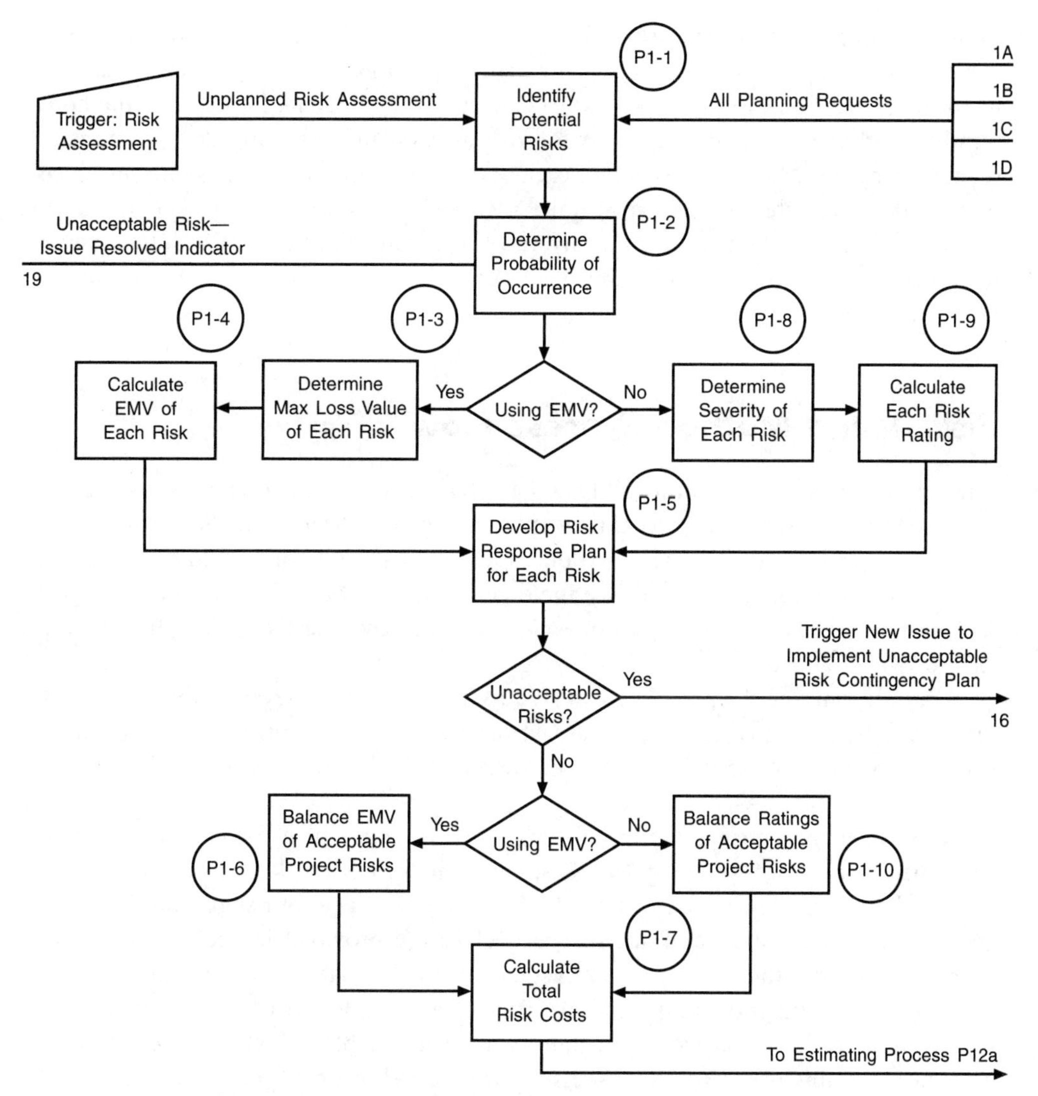

Figure 6.1 Risk Management Planning (P1) Process Flow

- An outstanding project event, such as unsatisfactory project performance or negative events in the foreseeable future
- Every time the project process flow enters the Planning Frame, which happens when the request for planning one of the project frames enters the Planning Frame via entry points 1A, 1B, 1C or 1D, as well as when a large change request is triggered

Process P1-1 will identify all potential risks, as described later in this chapter. When a list of project risks is assembled, the probability of each risk occurrence must be determined in the process Determine Probability of Occurrence (P1-2). It is necessary to rate the probability of each risk occurrence using a four-tier scale: **L** for low (less than 25%), **M** for medium (25 to 49%), **H** for high (50 to 74%) and **E** for exceptional (75 to 100%) (see the discussion of process P1-2 later in this chapter). In the EMV method, numbers rather than letters are used for risk assessment. Assigning an approximate medium range (10, 40, 65 and 90%) for each of the four ratings is usually sufficient, since it is almost impossible to rate a risk more accurately.

From this point, the two methods diverge. If the team is not using EMV, the answer to the decision point question (using EMV?) is no, and the next process executed is Determine Severity of Each Risk (P1-8).

Risk Rating Method Process Flow

After determination of risk probability and answering no to the decision point question (using EMV method?), the process flow enters the process Determine Severity of Each Risk (P1-8), where severity is assigned on a four-point scale: **L** for low, **M** for medium, **H** for high and **E** for exceptional (see the description of process P1-8 for details). Risk severity means the potential impact of each risk on project objectives, using the ranking scale provided above.

After determining the probability of risk occurrence and the severity, the overall risk ratings can be calculated in the process Calculate Each Risk Rating (P1-9). Risk ratings will be calculated on a four-level scale (L, M, H and E). The method is described in the section on process P1-9.

Now, when risk ratings are calculated, a risk response plan for each risk will be developed in the process Develop Risk Response Plan for Each Risk (P1-5), which is used in both the risk rating and EMV methods. There are four types of risk responses: avoidance, transfer, mitigation and acceptance. Details are provided in the description of process P1-5. After this process is complete, plans to eliminate or convert unacceptable risks (most medium and all high and exceptional risks) to low risks are developed. The process flow reaches a decision point question (unacceptable risks?). If there are unacceptable risks, this will trigger a new issue for each risk via exit point 16, which must be resolved in order to implement plans for elimination or reduction of those risks to low level. Some of those issues will generate scope change requests, as discussed in Chapter 10. Those plans must be implemented before going on with project execution. All exceptions to this rule must be specifically authorized by both delivery and business management after being warned of the danger of doing so. When an unacceptable risk is converted to an acceptable one during an issue resolution or scope change, the indication of the resolution comes back via entry point 19 to process P1-2 and the risk assessment process starts all over again.

When all risks have responses that make them acceptable, the process flow reaches a decision point question (using EMV?). Since the risk rating method is being used at the moment, the answer is no. Assuming that at this point only the low-rated risks are left, the overall cost of all acceptable low project risk remediation efforts will be calculated in the process Balance Ratings of Acceptable Project Risks (P1-10), as described in the corresponding section of this chapter.

The cost of risk remediation efforts, plus a margin (also known as a contingency reserve) for unknown risks will be added to project estimates, as described in the section on process P1-7. The process flow goes to the Estimating (P12a) process.

Expected Monetary Value Method Process Flow

EMV is a statistical assessment of project losses due to risks. While it is not possible to predict accurately the final cost of risks, this method is more accurate than the risk rating method, because instead of four levels of severity, the EMV method actually calculates the monetary value of risks in monetary units (U.S. dollars, euros, etc.). However, in order to achieve that accuracy, this method requires more effort and is more difficult to implement. Before selecting the best option, decision tree analysis may be used to analyze the most complicated risks in large projects with many possible remediation options and branch points. (An example of decision tree analysis can be found at www.mindtools.com/dectree.html.)

After execution of processes P1-1 and P1-2 is completed, which are used in both the risk rating and EMV methods, all risks are identified and the probability of their occurrence is expressed as a percentage.

The process flow now enters the process Determine Maximum Loss Value of Each Risk (P1-3). Here the maximum loss value of each risk is calculated in monetary units (see the description of process P1-3 below). The next process executed is Calculate EMV of Each Risk (P1-4) (see the description of process P1-4 below). The project flow enters the process Develop Risk Response Plan for Each Risk (P1-5), which is used in both the risk rating and EMV methods, where plans are developed for each risk identified. The risk rating method is repeated to the point where a decision point question (using EMV?) is encountered. Since the answer is yes and assuming that by now only the low risks are left, the overall cost of all acceptable low-risk remediation efforts will be calculated in the process Balance EMV of Acceptable Project Risks (P1-6), as described in the corresponding section.

Those costs of risk remediation efforts, plus a margin for unknown risks, will be added to project estimates, as described in the discussion of process P1-7. The process flow goes to the Estimating (P12a) process via exit point 20.

Identify Potential Risks (P1-1)

This process is used in both the risk rating and EMV methods. There are known fixed lists of project risks in every business field. Your organization likely will even have templates provided by a project management office. Using those templates, and taking advantage of past project wisdom, will greatly simplify risk assessment. In addition, there will be other risks specific to each particular project, to be used with the templates.

The following is a basic list of risk *areas*, or categories, which may be used for small to medium projects in the field of software development and electrical/electronic and mechanical engineering:

- Business requirements
- Project complexity
- Nonfunctional requirements
- Delivery team experience with similar projects
- Delivery time frame
- Project estimating
- Monetary exchange rates
- Environmental regulations
- Resources
- Client participation
- Relationship with client
- Contract elements
- Contract type
- Supplier or contractor issues
- Quality management processes
- Cost structure
- Pricing practices
- Integration with other projects
- Project management and established project management processes

Each area contains at least one risk, but may also contain several definitions of that risk, one for each level of severity. The definitions applicable to your project should be selected. For example, the business requirements area risk (before the project Planning Frame starts) may have identified the following risk definitions:

1. *Business requirements are completely documented and signed off on. Very few changes are expected.* This is obviously a low-severity risk (scenario 1).
2. *Business requirements are only partly documented. Moderate level of changes may be expected.* This is a medium-severity risk (scenario 2).
3. *Business requirements are not documented or not signed off on and are expected to be unstable.* This is the highest possible risk severity, because the lack of documented business requirements puts the successful implementation of the project in doubt. This is a high-severity risk (scenario 3).

As another example, the risk area project complexity may have three levels of risk severity:

1. *The solution or service is simple, with no customization and no interface requirements.* This is a low-severity risk (scenario 1).
2. *The solution or service has moderate complexity and requires modification or has one or two interfaces to other applications.* This is a medium-severity risk (scenario 2).
3. *The solution is complex and requires customer-specific functions or has more than two interfaces.* This is a high-severity risk (scenario 3).

For the construction business, a good source on risk assessment is the *Project Risk Management Handbook* (Office of Statewide Project Management Improvement [OSPMI], 2007, www.dot.ca.gov/hq/projmgmt).

In addition to predefined risks, there certainly will be other risks in every project, which must be taken into consideration. In addition to identifying a risk, it is always a good idea to have several possible definitions of it for different severities, as shown in the above examples. Remember to also look at your assumptions—these are often a rich source of risk identification.

Determine Probability of Occurrence (P1-2)

This process is used in both the risk rating and EMV methods. In the examples above, using the risk rating method, the probability of occurrence of the business requirements risk is low (L) for scenario 1, medium (M) for scenario 2 and high (H) for scenario 3. This corresponds to an average occurrence of 5, 20 and 40% for the EMV method, but the interim numbers are not used. When using the EMV method, the probability after the analysis is more accurate (see Table 6.1).

However, it is not always possible to provide several definitions for one risk. For example, relying on certain material or technology which is not yet available but is promised to be available on time may not be a good idea, unless the statistics for that specific supplier confirm their reliability. If the supplier keeps their promise 40% of the time and there are significant verifiable statistics proving that, the chance of the supplier failing to deliver on time is around 60%. The risk may be late delivery by the supplier, and the probability of its occurrence is 60% for the EMV method, which corresponds to exceptional (E) probability of its occurrence in the risk rating method.

NOTE When identifying risks, we recommend that you use a risk statement with an if-then format that shows the effect on project objectives. For example, instead of just listing supplier delays as a risk, it is better to more precisely communicate the risk as follows: If the supplier fails to deliver the server on time, this will delay the beginning of the coding process, affecting our ability to meet the client's schedule.

Table 6.1 Risk Probability Quantification

Probability of Risk Occurrence	
Risk Rating Method	**EMV Method**
Low (5%)	0–10%
Medium (20%)	11–30%
High (40%)	31–50%
Exceptional (75%)	51–100%

Another method, which is somewhat similar, lists several options for each risk (as used in the risk assessment tool downloadable free from www.pm-workflow.com). For example, the business requirements document is always a risk. The following options can be used to determine the level of risk:

1. Business requirements document is signed off on (low risk)
2. Business requirements document is available but not signed off on by the deadline (high risk)
3. Business requirements document is not available and requirements are not complete by the deadline (exceptional risk)

Determine Maximum Loss Value of Each Risk (P1-3)

This process is used in the EMV method. As the name implies, this is the maximum cost of a risk if it occurs or, in other words, the severity of the risk in monetary form. This number depends not only on the risk, but also on the whole project. If the project cost is $1 million and the maximum penalty for failing to deliver the project on time is also $1 million, then in the worst case scenario, when business requirements are not documented or not signed off on but estimates are finalized, the maximum loss value may be as much as $2 million or even more. Obviously, when the risk assessment is performed in accordance with the risk management process, the risk would never be accepted and this situation will not occur. If the project follows all processes presented in this book, the only acceptable risk scenario would be as follows: Business requirements are agreed to between the delivery team and all business teams, and all details are documented and signed off on. The quality management process will prevent the project manager from going any further in the project if the business requirements document is not signed off on, with notable exceptions, in which case the client takes responsibility for any business requirements issues. Therefore, the business requirements risk will be a low-severity risk and the maximum loss value will be in the range of 0 to 10% of the project cost, or in this case between $0 and $100,000. The EMV calculations will show later that the overall cost of risk will also depend on the probability of the risk occurrence. The cost of the late introduction of scope changes to the signed-off business requirements will be covered in Chapter 10.

The maximum loss value of the above-mentioned late delivery risk must take into consideration the cost of the workaround or the cost of the alternative, including a possible redesign, as well as the additional time required to deliver the solution.

Determine Severity of Each Risk (P1-8)

This process is used in the risk rating method. As mentioned at the beginning of the process flow section, severity is assigned as L, M, H or E. The overall guideline is:

- **L** Assigned when the maximum cost of a risk is less than 10% of the project cost
- **M** Assigned when the maximum cost of a risk is between 11 and 30% of the project cost
- **H** Assigned when the maximum cost of a risk is between 31 and 50% of the project cost
- **E** Assigned when the maximum cost of a risk is over 51% of the project cost

Calculate EMV of Each Risk (P1-4)

This process is used in the EMV method. Here, the EMV of every risk is calculated using the following formula:

$$EMV = Probability * MaxLossValue$$

Probability in this formula is the probability of risk occurrence, expressed as a percentage. MaxLossValue is the maximum loss to the project in monetary units if the risk occurs.

Using the previous examples for business requirements risk, the worst case scenario will have a probability between 50 and 100%, or 0.5 to 1.0, and the MaxLossValue is $2 million. The risk EMV will be between 0.5 * $2 million = $1 million and 1.0 * $2 million = $2 million.

NOTE A probability of 100% would mean, for example, expecting on-time delivery from a vendor when the vendor was selected, but the vendor has since declared bankruptcy and there is no chance of getting any delivery.

If all processes are followed, then the probability of occurrence of the business requirements risk will be low or less than 10% when the MaxLossValue at low severity is $50,000 (see the description of process P1-3). The risk EMV will be 0.1 * $50,000 = $5,000.

Calculate Each Risk Rating (P1-9)

This process is used in the risk rating method. There is no specific formula that can be used, but the empiric method offered below allows selection of the overall risk rating based on the probability and severity of the risk. The intersection of probability and severity ratings will indicate the overall risk rating. The risk rating method is illustrated in Table 6.2. The alternate way to determine the overall risk rating is displayed in Table 6.3. The total project risk rating calculation and the corresponding costs are described below in the description of process P1-10.

Table 6.2 Risk Rating Method

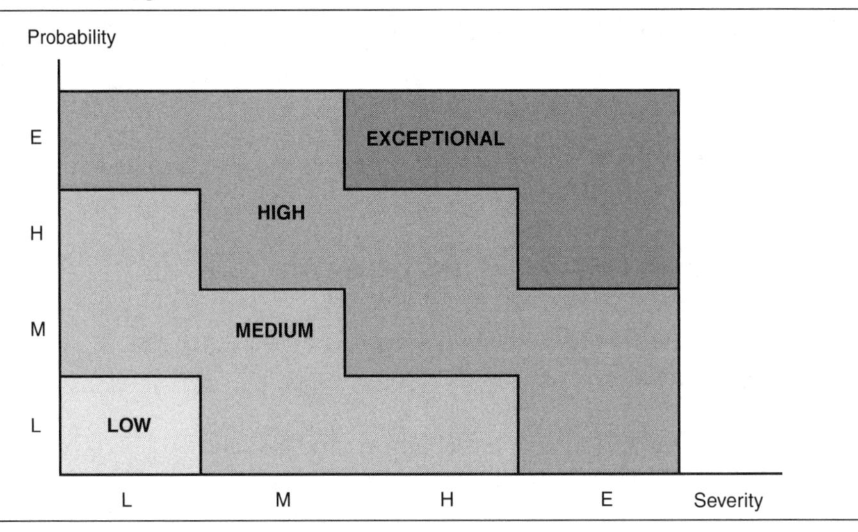

Table 6.3 Alternate Risk Rating Method

Probability	L	L	L	L	M	M	M	M	H	H	H	H	E	E	E	E
Severity	L	M	H	E	L	M	H	E	L	M	H	E	L	M	H	E
Risk rating	**L**	**M**	**M**	**H**	**M**	**M**	**H**	**H**	**M**	**H**	**H**	**E**	**H**	**H**	**E**	**E**

Using Monte Carlo Analysis to Assess Project Risk

Although a detailed review of this technique is outside the scope of this book, we felt that at least a basic understanding of this technique would be helpful. Monte Carlo analysis is used in commercial risk assessment tools and in very large mechanical, aeronautical, military and other projects. It is rarely used in electronics and software daily project activities due to its high overhead.

Monte Carlo analysis involves determining the impact of the identified risks by running computer simulations (repeated calculations of all aspects of the project involving uncertainty) to find a range of possible outcomes for various scenarios. A random sampling is performed by using uncertain risk variable inputs to generate these outcome ranges, with each outcome given a level of confidence. This is typically done by establishing a mathematical model and then running simulations based on this model to estimate the impact of project risks. This technique helps in forecasting the likely outcome of an event and thereby helps in making informed project decisions.

Using worst case, most likely, and best case expected values and the probability distribution curve associated with the task or attribute (complete system test), a project manager can use Monte Carlo analysis simulations to generate the most likely outcome for the event. In most situations, the normal distribution pattern for the possible outcomes is displayed as a bell-shaped curve. In the simple graphical form in Figure 6.2, the output of a Monte Carlo analysis shows cumulative probability and the actual samples plotted on the same chart. From this configuration, a project manager could, for ex-

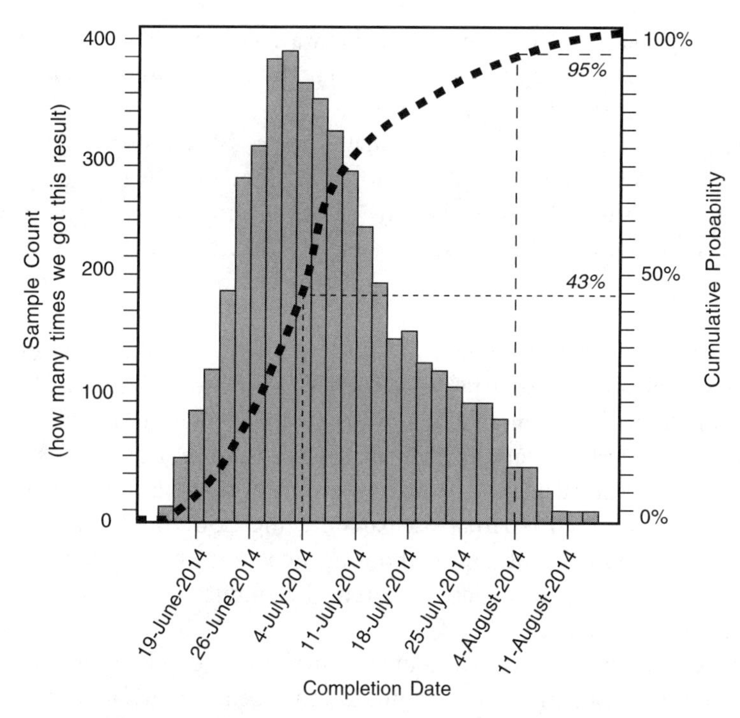

Figure 6.2 Monte Carlo Risk Assessment Analysis

ample, give a sponsor a completion date or budget with a 50, 70 or 90% level of confidence or, alternatively, determine the confidence level for meeting a certain completion date or budget.

In the example in Figure 6.2, the cumulative probability is plotted with a dotted line, in a continuously increasing percentage from 0 to 100% as more and more of the samples (which are shown as grey bars) are included. Based on this, the project manager could inform the sponsor that the chance of completion by July 4, 2014 is 43% or the completion date with a 95% confidence level is August 4, 2014.

Develop Risk Response Plan for Each Risk (P1-5)

The risk response plan is not the same as the risk management plan. The risk management plan details how in general risk in the project will be dealt with. In practical terms, the risk management plan is a schedule of planned risk assessments and a description of when and under what conditions the risk assessments take place. In contrast, the risk response plan deals with the specific risks identified in each project risk assessment.

This process is used in both the risk rating and EMV methods. There are four general risk response approaches, as explained in the following paragraphs.

Avoidance. Use a different approach to project implementation when the impact of a risk on the project is too high and therefore the risk is unacceptable. This approach will eliminate exceptional, high and most medium risks or reduce them to the acceptable low level. To avoid the risk, a different approach to project implementation should be used. Here, the project plan is changed in some way. This may be very costly, especially when the project is in the later stages of implementation. The cost of risk avoidance must be estimated for the overall risk cost calculation.

Transfer. This is another way to eliminate high risks by transferring responsibility for implementation of some project elements to a different delivery group or contracting them out. This is usually done when the delivery team does not have the experience or human resources to implement some project elements. It is, in essence, a form of insurance. The cost of transferring risks must be estimated for the overall risk cost calculation.

Mitigation. Develop and implement plans to reduce the likelihood of risk occurrence or its severity. The overall rating of the risk after mitigation must be reduced to low (for example, use older methods of project implementation or older materials instead of unproven latest methods or materials). The cost of mitigation must be estimated for the overall risk cost calculation. A good example of this is a project which has as its outcome an outdoor event (perhaps a party). Mitigation could involve providing a canopy to reduce the *impact* of rain or choosing a date on which historically the chance of precipitation is lower, to (at least theoretically) reduce the *probability* of rain. Of course, both could be done.

Acceptance. Accept the risk without any attempt to eliminate it or to reduce its rating. Active acceptance means developing a contingency plan for each accepted risk. The plan will be implemented only if the risk occurs. Passive acceptance means having no contingency plan. Usually, most low risks and some medium risks are accepted when losses due to their occurrence are lower than the cost of mitigating them. If those risks are triggered during the course of the project, they will cause predictable losses, which are equivalent to the cost of the contingency plan implementation, if one was created. Keep in mind that acceptance without a contingency plan means the cost of contingency implementation if the risk occurs is unknown. Passive acceptance should be applied only to low risks. Otherwise, the unknown cost of mitigation may present a high risk. All options should be analyzed and the most cost-effective method of risk responses used.

Balance EMV of Acceptable Project Risks (P1-6)

This process is used for risk quantification in the EMV method. In this process, costs of all acceptable risks are summarized. As shown in Table 6.4, assuming that the total project cost is $1 million, $27,700 must be added to the detailed project estimates as the acceptable risks margin.

Balance Ratings of Acceptable Project Risks (P1-10)

This process is used for risk quantification in the risk rating method, as shown in Table 6.5. In this process, costs of all acceptable risks are summarized as shown in the following example.

Table 6.4 Risk Quantification Example Using EMV Method

#	Risk Name	Probability	MaxLossValue	Risk EMV
1	Risk #1	9%	$60,000	$5,400
2	Risk #2	7%	$40,000	$2,800
3	Risk #3	15%	$70,000	$10,500
4	Risk #4	10%	$90,000	$9,000
	Total risk costs (sum of all risk EMV)			$27,700
	Total risk costs as % of project cost			2.8% (L)

Table 6.5 Risk Quantification Example Using Risk Rating Method

#	Risk Name	Probability	Severity	Overall Rating
1	Risk #1	L	L	L
2	Risk #2	L	L	L
3	Risk #3	M	L	M
4	Risk #4	M	L	M
	Normalized risk score (see below)			35 (L)
	Total risk cost for low risk score (35) is <5% of project cost ($1 million)			<$50,000

The total risk cost calculation is done using the normalized risk score method:

1. Rate each risk and add up all the risk scores—The risk score is rated as follows: 2 (low), 5 (medium), 9 (high), 14 (exceptional). For example:

$$\text{Total risk score} = 2 + 2 + 5 + 5 = 14$$

2. Normalized score = Total risk score * (10/Number of risks)

$$\text{Normalized score} = 14 * (10/4) = 35$$

3. Overall risk ratings: 21 to 40 = low, 41 to 60 = medium, 61 to 80 = high, 80+ = exceptional. A normalized score of 35 in this example corresponds to an overall low risk.

L Low, which corresponds to a maximum cost of risk less than 5% of the project cost

M Medium, which corresponds to a maximum cost of risk between 5 and 20% of the project cost

H High, which corresponds to a maximum cost of risk between 21 and 50% of the project cost

E Exceptional, which corresponds to a maximum cost of risk over 51% of the project cost

Since the overall low risk costs less than 5% of the total project cost, then the accepted risk costs in this project are less than $50,000. This is the highest accuracy that can be achieved using the risk rating method.

Calculate Total Risk Costs (P1-7)

This process is used in both the risk rating and EMV methods. Total risk costs are the sum of the following costs:

1. Accepted risks
2. Avoided risks
3. Transferred risks
4. Mitigated risks
5. Unknown risks

Unknown risks are risks that are beyond existing knowledge. They cannot be assessed and no remediation plans can be built. Unknown risks may come from any area over which the project manager does not have control or may be due to changing circumstances in the course of the project. The best way to estimate unknown risks is to examine statistics for the past several years. If statistics show the cost of unknown risks was 40% of the cost of known risks, the same cost margin may be used, since repeated failures are almost guaranteed. If no statistics are available, this signals the lack of project management processes or poor quality management. The best thing that a project manager can do in this situation is insist on strictly following those processes. If the current organizational culture allows implementation of those processes on the enterprise level, assume the unknown risks cost margin at 20% of the cost of all other risks. Otherwise, repeated failures are guaranteed not only in the risk management area, but in all project management areas. This by itself can put a project in the exceptional risk area. It is most probable that management in such an organization will not allow inclusion of high risks associated with lack of implementation of quality assurance and other project management processes, including risk assessment. In some enterprises, the "unknown unknowns" are covered across projects with a management reserve, to handle, for example, a labor action which would affect an entire portfolio of projects. Unlike the contingency reserve mentioned earlier, which is under the control of the project manager, the management reserve is under the control of a senior manager in the enterprise.

Risk Management Plan

The risk management plan provides guidance on when and how, in general, to manage risks throughout the project, as described above. The schedule of risk management activities is built into the existing project plan. The risk management plan includes the following topics:

- Risk management objectives
- Risk management processes
- Scales for probability and severity
- Established schedule and frequency of risk assessment
- Risk management roles and responsibilities
- Reporting risk issues

It may not be necessary to have a risk management plan as a separate document, because all the plan topics already have been laid out in the preceding sections. Risk management processes, lists of all risks that occurred, new risks found and results of risk assessments should have been documented in the project control book. The schedule of risk assessment should have been embedded in the existing project plan. Results of risk assessments are reported to senior delivery management at least monthly.

Risk Metrics

Risk metrics that can be used by the project team include:

- Record of each risk assessment and the reason for assessment
- Total number of risk assessments to date
- Total number of planned risks that occurred (this comes from the issue management metrics)
- Total number of unplanned risks that occurred (this comes from the issue management metrics)
- Planned cost of each risk mitigation (before the risk occurs) and the overall planned cost of risk mitigation
- Actual cost of each risk mitigation and the overall actual cost of risk mitigation
- Planned cost of each risk that occurred and the overall planned cost of risks that occurred (this comes from the issue management and the scope change control metrics)
- Actual cost of each risk that occurred and the overall actual cost of risks that occurred (this comes from the issue management and the scope change control metrics)

Risk Assessment Form

Table 6.6 shows a risk assessment form.

Project Risk Assessment Tool

A risk assessment tool for small to medium projects in the engineering area is available for download from J. Ross Publishing (from the Web Added Value™ Download Re-

Table 6.6 Risk Assessment Form

Project Risk Assessment
Project Name: ______ Risk Assessment Number: ______ Date: ______ Total Acceptable Risks Score/EMV: ______ Calculation of the Total Acceptable Risks Score/EMV (or reference to a separate document): ______ ______ Total Cost of All Risk Remediation Plans: ______
Risk Information
Risk Number: ______ Risk Owner: ______ Risk Identification: ______ Probability and Justification of Probability (or reference to a separate document): ______ ______ Severity/EMV and Justification of Severity/EMV (or reference to a separate document): ______ ______ Risk Score/EMV and Calculation of Risk Score/EMV (or reference to a separate document): ______ ______ Risk Remediation Method: ☐ Avoidance ☐ Transfer ☐ Mitigation ☐ Acceptance Risk Remediation Plan (or reference to a separate document): ______ ______ Avoidance/Transfer/Mitigation Plan Due Date: ______ Remediation Plan Implementation Estimates (or reference to a separate document): ______ ______
Risk Information
Enter next risk information here:

source Center at www.jrosspub.com), as well as from the authors (at www.pm-workflow.com). The tool is applicable for the risk score method of risk assessment. The entry page of the tool, as shown in Figure 6.3A, requires entering the customer organization name, the project name and the date of the risk assessment, which are propagated to other assessment pages. The first risk assessment page is displayed in Figure 6.3B.

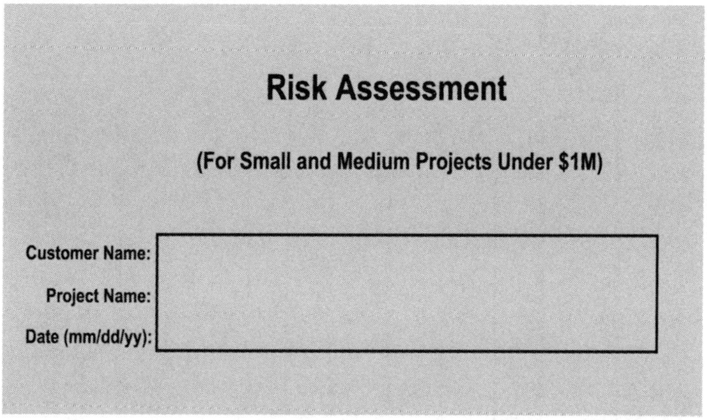

Risk Assessment

(For Small and Medium Projects Under $1M)

Customer Name:
Project Name:
Date (mm/dd/yy):

Figure 6.3A Screenshot of Entry Page of the Risk Assessment Tool

There are several pages in the tool, with 20 standard risk groups listed. Each risk is presented with several possible ways it may be represented. For example, the first risk group, business requirements, has four options:

1. Not applicable
2. Business requirements document is signed off by due date
3. Business requirements document is available, but not signed off yet by due date
4. Business requirements document is not available and requirements are not complete by due date

When a selection is clicked on, the risk score is displayed to the right of the risk definition. The risk score shown is a derivative of risk probability and its severity. For risks that have medium, high and exceptional scores, there is a risk remediation suggestion in the tool, which is the basic risk remediation plan. For example, if option #3 is selected, the letter M for medium risk score is displayed and a list of three suggested risk remediation actions is displayed (in red):

- *Ensure that business requirements are signed off by all parties and estimates are agreed*
- *Ensure that a project charter is received and a kickoff meeting has taken a place*
- *Ensure all project management processes are established and mandatory*

If all three suggestions are implemented, the business requirements risk will change from medium to low. All 20 standard risks must be assessed and the resulting risk scores for standard risks will be automatically calculated.

For unique risks, the risk description must be entered in the appropriate screen in the risk identification section and both severity and probability must be selected, as displayed in Figure 6.3C. The resulting risk score will be displayed on the right of the screen.

Common Risk Statements 1-6

Customer Name:
Project Name:
Date (mm/dd/yy):

Score: E - Exceptional, H - High, M - Medium, L - Low

Risk Identification	Select Risk	Risk Score
1. Business Requirements		
<N/A> Not Applicable	○	
<L> Business Requirements Document is signed off by due date.	○	
<M> Business Requirements Document available, but not signed off yet by due date	●	M
Ensure that business requirements are signed off by all parties and estimates are agreed.		
Ensure that a project charter is received and a kick off meeting has taken place		
Ensure all PM processes are established and mandatory.		
<H> Business Requirements Document is not available and requirements are not complete by due date	○	
Produce business requirements and have it signed. Ensure that estimates are produced and agreed by all parties		
Ensure that a project charter is received and a kick off meeting has taken a place		
Ensure that all PM processes are enforced		
2. The delivery team familiarity with similar projects		
<L> The Delivery Team is familiar with similar projects.	●	L

Figure 6.3B Screenshot of First Risk Assessment Page for Common Risks

Risk Statements - Unique Risks

Customer Name: John Smith

Project Name: Venus Vehicle

Date (mm/dd/yy): 1/15/2013

Risk Identification	Severity of Risk if Occurs	Probability of Occurence	Resulting Risk Score
Risk Description *Risk Remediation*	○ Except ○ High ○ Medium ○ Low ◉ N/A	○ Except ○ High ○ Medium ○ Low ◉ N/A	N/A
Risk Description *Risk Remediation*	○ Except ○ High ○ Medium ○ Low ◉ N/A	○ Except ○ High ○ Medium ○ Low ◉ N/A	N/A
Risk Description *Risk Remediation*	○ Except ○ High ○ Medium ○ Low ◉ N/A	○ Except ○ High ○ Medium ○ Low ◉ N/A	N/A

Figure 6.3C Screenshot of Risk Assessment Page for Unique Risks

Risk Summary

Customer Name: Customer1
Project Name: Project1
Date: mm/dd/yyyy

Copyright © Daniel Epstein

Known Risks	# of Risks	Normalized Risk Score	Overall Risk
TOTAL PROJECT	**17**	**25.9**	**LOW**
Exceptional	0		
High	1		
Medium	1		
Low	15		

Figure 6.3D Screenshot of Risk Tool Summary

When the risk assessment is complete, the last screen will display a risk summary. The total number of risks will be displayed, along with separate counts for low, medium, high and exceptional risks, as shown in Figure 6.3D. The overall normalized risk score for the entire risk assessment will be displayed, along with the overall project risk rating. The total project cost and the method described above can be used to convert the overall risk score into monetary value.

Work Breakdown Structure Design and Preliminary Project Planning (P5)

Purpose

The purpose of the work breakdown structure (WBS) is to establish plans and methods for implementing and managing projects. Recall from Chapter 1 the definition of a WBS as used in the context of this book. The WBS is a tool for developing or updating the schedule data, such as milestones, deliverables, dependencies, risks, work products and resource requirements. Changes to the project scope any time throughout the project will bring the project flow back to this process. The WBS is a foundation upon which task estimates, task dependencies, resource allocation, risk management, quality management and project planning are built into the project schedule.

Work Breakdown Structure Activity Decomposition

The WBS is a deliverable-oriented multilevel decomposition of the entire project scope into sets of smaller, more manageable and controllable tasks. The WBS is, in effect, a hierarchical tree, which may be presented in several different forms:

- WBS decomposition diagram (hierarchical)
- Gantt chart (adds element of time)
- Network diagram (focused on sequencing and dependencies)

WBS decomposition rules are different from process decomposition rules, which were described in the Requirements Frame section. Instead of decomposing high-level business activities into elementary business activities during process decomposition, WBS activities are decomposed down to the level of the smallest executable project element with an identified deliverable, called a task. A deliverable does not have to be a client deliverable; it can be an intermediate deliverable, which is required to create a client deliverable. There are two important rules of WBS decomposition:

- No task should take longer than 40 hours (since any longer would indicate that further decomposition is possible).
- No task should take less than 16 hours (since this could cause an excessive number of small tasks to track).

The 40-hour rule allows the project manager to have better control over the project. Experienced project managers know that when they ask team members about the status of not yet completed tasks, the answer is often 80 or 90% completion. Unfortunately, the remaining 10 or 20% takes three times longer than the previous 80 or 90%. In fact, it is impossible to know the real status of a task until it is 100% complete and the deliverable is available for review. Using this rule, the worst delay in task completion would be 40 hours, after which corrective measures can be taken. If a task takes 8 weeks, it will become obvious 8 weeks down the road that the task is far from complete.

The correctly designed WBS for a medium-size project has between several hundred and a thousand tasks. If many tasks take less than 16 hours, then the WBS for the project can have thousands of tasks, which makes the project less manageable. There is an exception for mini-projects with durations of several days or weeks, which may have shorter tasks. WBS creation should involve working sessions with all key technical personnel.

A decomposition diagram for a child's birthday party is presented in Figure 7.1. The diagram is just a sample, and many essential tasks are not included for the sake of keeping it simple. All activities and tasks are shown as rectangles. The top-level activity, Child Birthday Party, is decomposed into four activities:

- Invite Guests
- Get Supplies
- Conduct Party
- Clean the Mess

A decomposition diagram is not a chronological list of tasks. At this point, sequencing and duration of tasks are not really considered. The focus of the WBS at this point is clearly delineating what work the project involves. The WBS is a complete list of tasks

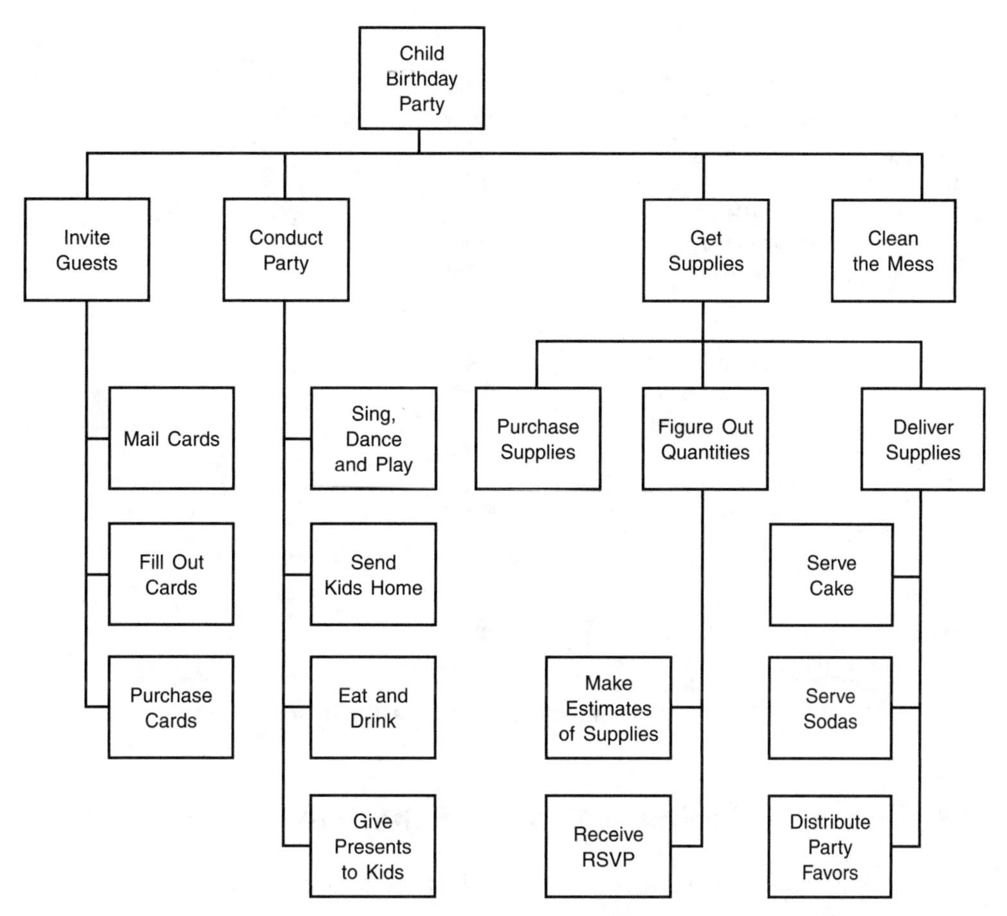

Figure 7.1 Activity Decomposition Diagram

required to deliver the entire project scope with no regard to their sequence, duration or assigned resources yet. Each task must point to the specific element of the project scope. While more than one task may be needed to implement one element of the project scope, there should be no duplication of effort, where two tasks do the same work.

The activity decomposition diagram in Figure 7.1 represents a list of tasks required to deliver the mini-project. If the WBS has several hundred tasks, it is impractical to draw this type of diagram.

The WBS is now used as the indented task list on the left-hand side of a typical Gantt chart (see Figure 7.2). This list, as we continue working with the WBS, gradually evolves into the schedule with the addition of task duration and sequencing information. The Gantt chart is used in all project scheduling tools, such as Microsoft Project, Clarity, Primavera and others. The lowest level executable activity is called a task. All others are called activities and are presented in the Gantt chart mostly for clarity and context. Higher level (or "rolled-up") activities are displayed in bold type, while subordinate individual task names are in regular type, as shown in Figure 7.2.

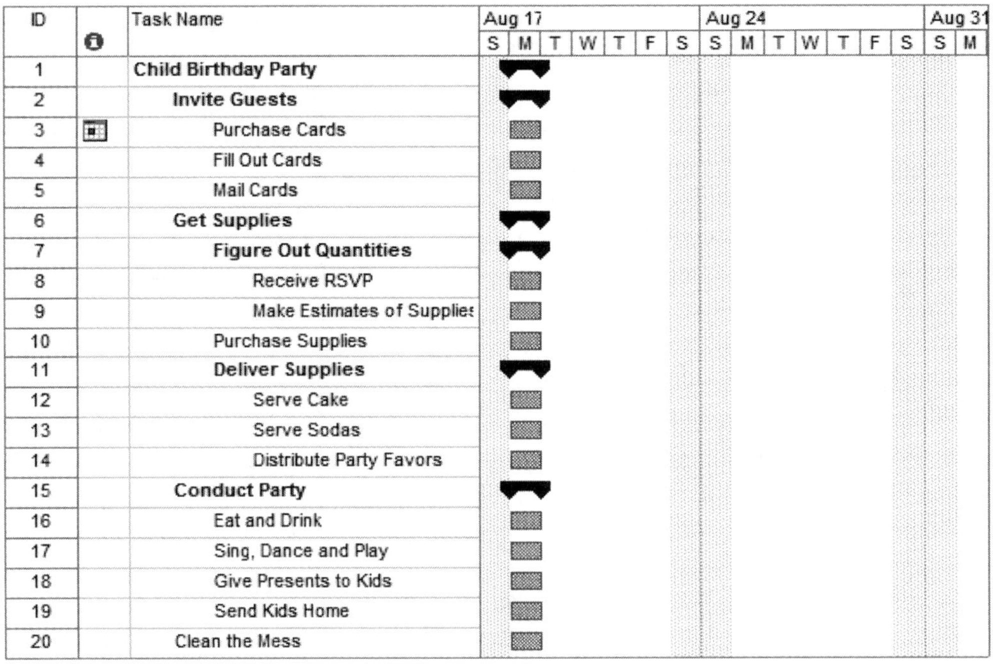

ID		Task Name
1		**Child Birthday Party**
2		**Invite Guests**
3		Purchase Cards
4		Fill Out Cards
5		Mail Cards
6		**Get Supplies**
7		**Figure Out Quantities**
8		Receive RSVP
9		Make Estimates of Supplie:
10		Purchase Supplies
11		**Deliver Supplies**
12		Serve Cake
13		Serve Sodas
14		Distribute Party Favors
15		**Conduct Party**
16		Eat and Drink
17		Sing, Dance and Play
18		Give Presents to Kids
19		Send Kids Home
20		Clean the Mess

Aug 17 S M T W T F S | Aug 24 S M T W T F S | Aug 31 S M

Figure 7.2 Indented Task List for a Gantt Chart

Building a Preliminary Project Schedule

The indented task list is the first step in building the project schedule. To develop it further, the following steps are required:

1. Enter task dependencies (also called task precedence)
2. Estimate all tasks
3. Assign resources

Building Task Dependencies

Task dependencies indicate relationships between tasks. For example, you cannot mail out cards (task) before you purchase them (task). Also, sometimes a delay is required between the end of one task and the beginning of another, as in the case of mailing cards and receiving RSVPs. The dependency relationship is shown in Figure 7.3. Tasks may be taken on in one of the following relationships:

- Finish to start (task starts at the time when a predecessor task ends)
- Start to start (task starts at the time when a predecessor task starts)
- Finish to finish (task ends at the time when a predecessor task ends)
- Start to finish (task ends at the time when a predecessor task starts)

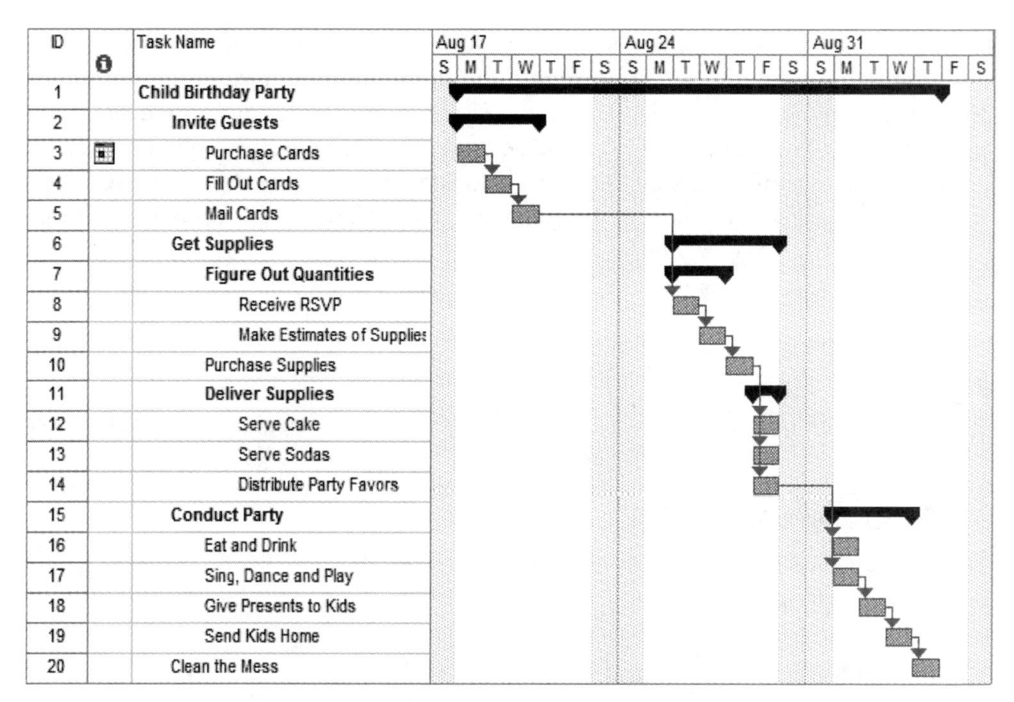

Figure 7.3 Task List with Dependencies

By far, the most commonly encountered sequence of tasks is finish to start, where one task starts after the previous task ends. In each of the above relationships, the lag between each task and its predecessor is 0 by default, but may be defined differently. In our case, a lag of 5 days is the waiting time required between the finish of the task Mail Cards and the start of the task Receive RSVP.

Entering Task Constraint Type

Another important parameter that should be taken into consideration is constraint type, which describes the period when a task should start or finish. The most frequently used constraint type is as soon as possible, which means the next task should start immediately, unless a lag is defined. There are eight constraint types, which will be explained later, when critical path analysis is covered:

1. As soon as possible
2. As late as possible
3. Finish no earlier than
4. Finish no later than
5. Must finish on
6. Must start on
7. Start no earlier than
8. Start no later than

Constraint types 3 to 8 are useful, but must be used with extreme caution, because they are tied to specific dates. If the schedule is later modified, tasks with constraint types 3 through 8 will not move, which creates confusion in project scheduling and generates error messages by scheduling tools. Inexperienced project managers often make the mistake of using a must-start-on constraint and later are unable to update the schedule unless constraints are modified. In a large project with a thousand or more tasks, it can take some time to fix this problem.

A *must-start-on* constraint should be used for tasks like scheduled training with many participants, which are difficult to reschedule. The predecessor task was supposed to have been completed a while ago, so if the schedule changes, there is still enough time to start the task at the scheduled date.

A *must-finish-on* constraint is used when a task must be finished on a specific date.

Entering Task Estimates

The next step in building a schedule is estimating all tasks. Each task has two measurements:

- Effort
- Duration

Effort is the number of labor hours required to accomplish a given task. If a resource is assigned to work 100% on the task, which is called 100% availability or 100% load, then the task duration will be equal to the task effort. However, if resource availability is 50%, then twice as many workdays will be required to accomplish the same task. For example, if the task effort is estimated to be 20 hours and the resource availability is 50%, then the task duration will be 40 hours.

Milestones are mentioned many times in this book. As a reminder, a milestone is a task with a duration of 0 which serves as a flag or indication that certain parts of the project are complete. Some project managers like to think of milestones as “anchors” because they help to tie down the project schedule and help connect it to the outside world more directly. In our example, we have the following milestones:

- Cards Mailed
- Ready to Have Fun
- Thanks, It Is Over

After estimating task length, our Gantt chart looks like Figure 7.4.

Resource Assignment

The next step is to assign resources to the project. In the example in Figure 7.5, John’s availability is 25%, so the required effort of 2 hours for Purchase Supplies has a total

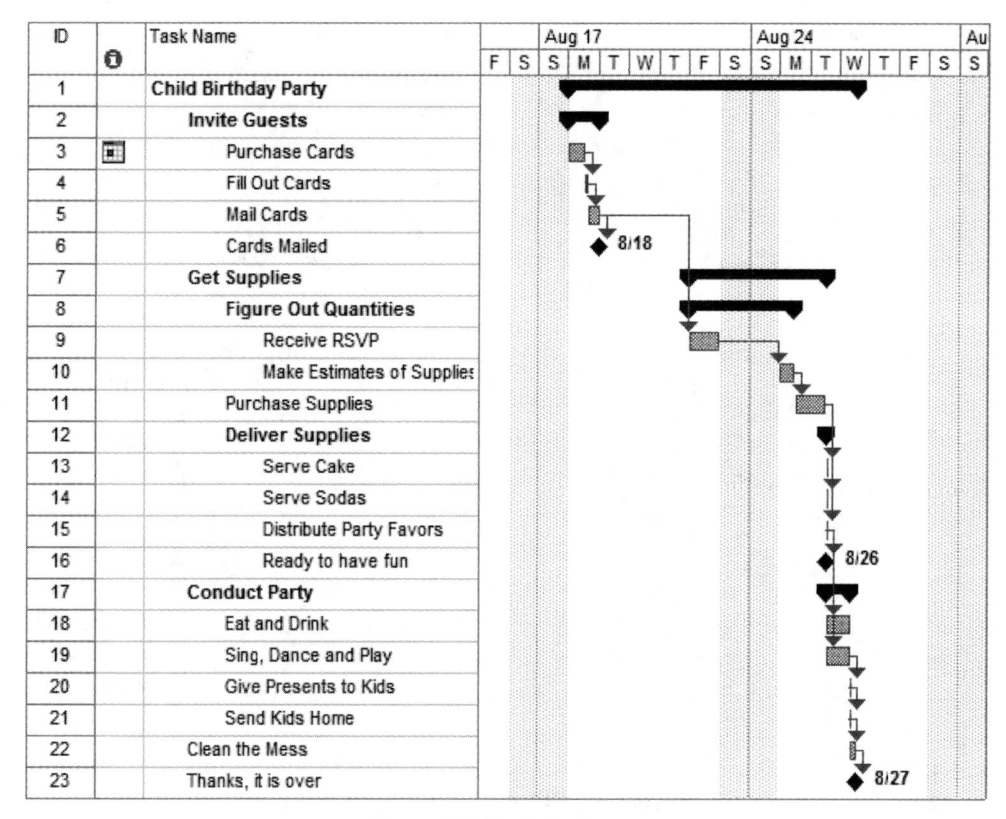

Figure 7.4 Task Estimates

duration of 8 hours. Also, tasks 18 and 19, which are performed by the resource Children, should take place at the same time. If that resource had 100% availability, the scheduling tool would automatically put one after the other, because a resource cannot be scheduled for 200% availability. In order to prevent this from happening, resource availability must be assigned at 50%. In total, the duration of each task is 4 hours, performed at the same time, while the effort devoted to each task is 2 hours. Resource assignment is displayed in Figure 7.5.

Modifying the Project Completion Date

It appears that the project will be completed on 8/27, which is when the party is scheduled to take place. But what if the child's birthday is on 8/25 and the party must be held on that date? One of many ways to accomplish this is by setting a must-start-on constraint on the first task, Purchase Cards, moving the end of the project back 2 working days to 8/25, as displayed in Figure 7.6. There are additional methods that may be used to accomplish this schedule compression. Those methods will be reviewed in the next section.

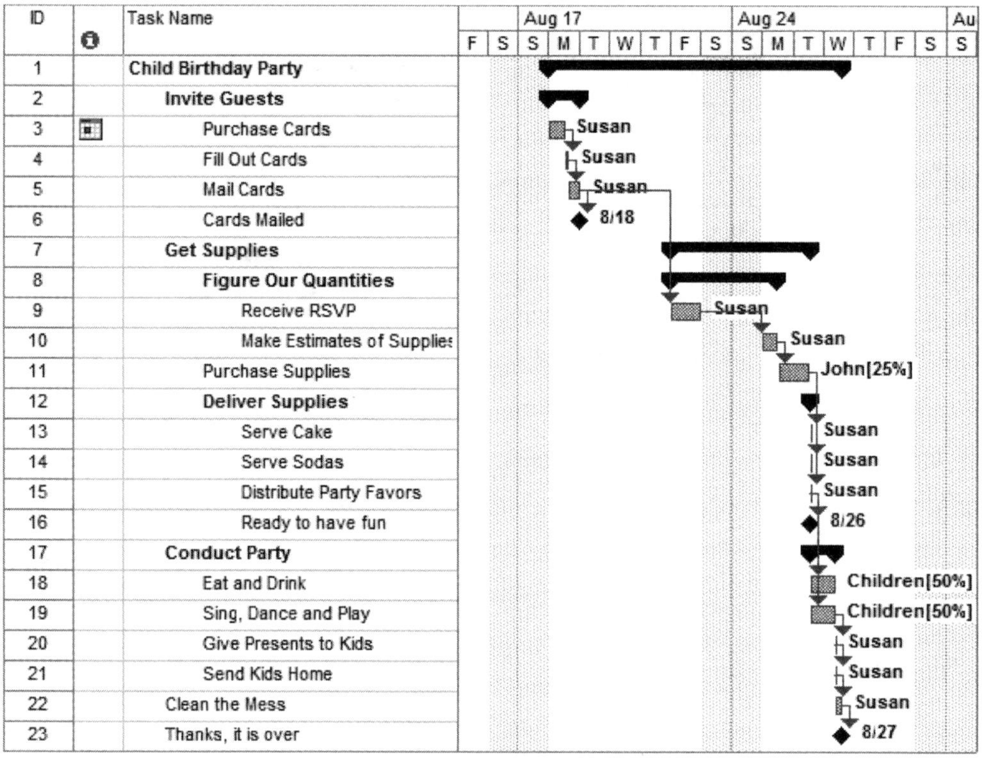
ID
Task Name
Aug 17
Aug 24
Au
F S S M T W T F S S M T W T F S S
1 Child Birthday Party
2 Invite Guests
3 Purchase Cards
4 Fill Out Cards
5 Mail Cards
6 Cards Mailed
7 Get Supplies
8 Figure Our Quantities
9 Receive RSVP
10 Make Estimates of Supplies
11 Purchase Supplies
12 Deliver Supplies
13 Serve Cake
14 Serve Sodas
15 Distribute Party Favors
16 Ready to have fun
17 Conduct Party
18 Eat and Drink
19 Sing, Dance and Play
20 Give Presents to Kids
21 Send Kids Home
22 Clean the Mess
23 Thanks, it is over
Susan
Susan
Susan
8/18
Susan
Susan
John[25%]
Susan
Susan
Susan
8/26
Children[50%]
Children[50%]
Susan
Susan
Susan
8/27

Figure 7.5 Resource Assignment

Critical Path Analysis

Critical path analysis is a method of identifying tasks that must be completed on time in order for the whole project to be completed on time. Those tasks are called critical tasks. Note that critical tasks are not necessarily any more "important" than other tasks. What makes them "critical" is that any slip in any critical path activities will, by definition, cause a slip in the overall project end date.

This means that noncritical tasks may have a calculated maximum allowable delay in their completion without delaying project completion. Critical path analysis calculates the allowable delay between the completion of one task and the beginning of the next one. It allows project managers to calculate project duration and to improve resource utilization.

The network diagram or PERT chart presents all tasks as small rectangles or circles. It shows tasks, their dependencies and their duration, just as a Gantt chart, but also helps to identify the longest path in the project, which is called the critical path. The birthday party project is too simple to show a meaningful network diagram; therefore, a generic and slightly more complex project is used in Figure 7.7 to illustrate the critical path analysis concept. This generic project has 15 tasks identified by numbers 1 through 15. The duration of each task in days is indicated on top of each rectangle.

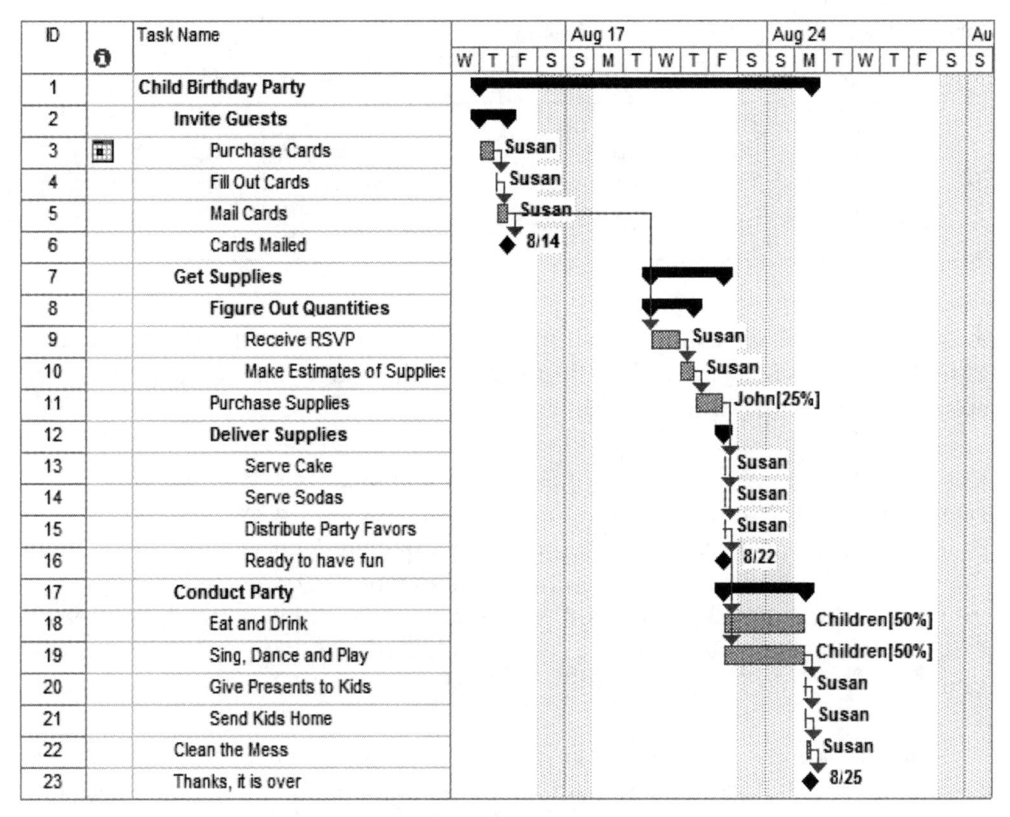

ID		Task Name
1		Child Birthday Party
2		Invite Guests
3		Purchase Cards
4		Fill Out Cards
5		Mail Cards
6		Cards Mailed
7		Get Supplies
8		Figure Out Quantities
9		Receive RSVP
10		Make Estimates of Supplies
11		Purchase Supplies
12		Deliver Supplies
13		Serve Cake
14		Serve Sodas
15		Distribute Party Favors
16		Ready to have fun
17		Conduct Party
18		Eat and Drink
19		Sing, Dance and Play
20		Give Presents to Kids
21		Send Kids Home
22		Clean the Mess
23		Thanks, it is over

Figure 7.6 Moving Completion Date Ahead

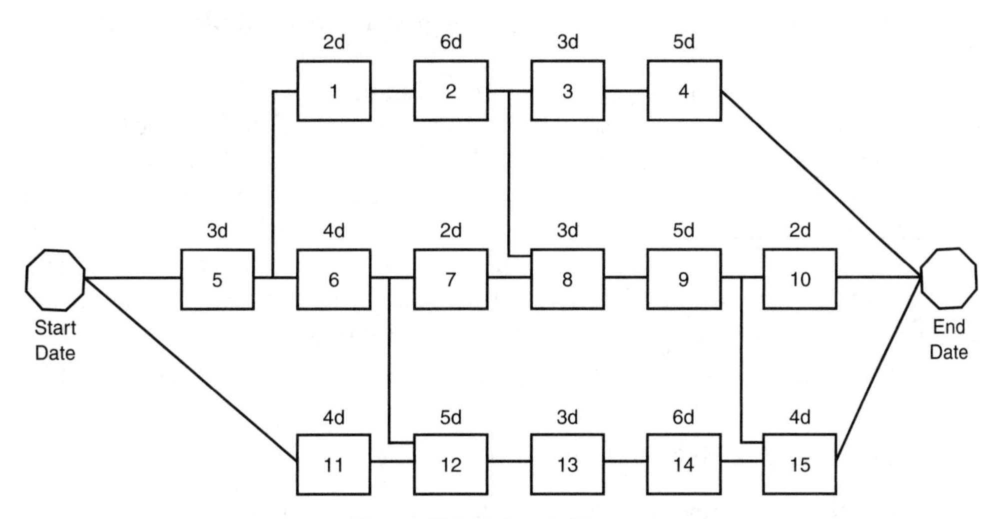

Figure 7.7. Network Diagram

Table 7.1 Task Start and Finish Date Example

Task	Start Date (morning)	Finish Date (end of the day)
5	01/01	01/03
1	01/04	01/05
2	01/06	01/11
3	01/12	01/14
4	01/15	01/19

The project path is a contiguous chain of tasks from the start to the end of a project. Note that technically "start" and "end" are tasks with a duration of 0. There are seven paths in the network diagram in Figure 7.7:

Path 1: Start, 5, 1, 2, 3, 4, End	Total duration: 19 days
Path 2: Start, 5, 1, 2, 8, 9, 10, End	Total duration: 21 days
Path 3: Start, 5, 1, 2, 8, 9, 15, End	Total duration: 23 days
Path 4: Start, 5, 6, 7, 8, 9, 10, End	Total duration: 19 days
Path 5: Start, 5, 6, 12, 13, 14, 15, End	**Total duration: 25 days** (the longest path)
Path 6: Start, 5, 6, 7, 8, 9, 15, End	Total duration: 21 days
Path 7: Start, 11, 12, 13, 14, 15, End	Total duration: 22 days

Note that it is not the number of tasks in a path that is important, but rather the cumulative (added) duration in the path. Path 5 in the above example is the critical path, because it is the longest path, which in turn establishes the project duration of 25 days. If the project start date is the morning of 01/01, then the project end will be the end of the day on 01/25, assuming that Saturdays and Sundays are workdays. If all tasks have an as-soon-as-possible constraint, then all tasks in path 1 start and finish on consecutive dates, as shown in Table 7.1.

Let's look at task 4. The earliest possible date it can start is 01/15, which is called the early start. Subsequently, the earliest possible finish date is 01/19, which is called the early finish. If the start date is delayed for 2 days, the task will be finished on 01/21 without impacting the project schedule, because 01/21 is still several days before the scheduled end of the project on 1/25. The latest date the task can finish without causing delay of the project is 01/25, which is called the late finish. Subsequently, the latest date the task can start is 01/21, which is called the late start. The difference between the late start and early start (or late finish and early finish) is 5 days, which is called the slack or float. This is the time by which the task can be delayed without impacting the project schedule.

Compressing the Schedule

During the course of a project, often the project or some tasks must be completed earlier than scheduled, due to new imposed dates or other significant issues. If the completion of tasks not on the critical path is accelerated, the project duration in our example still will be 25 days. It really does not make any difference if the duration of path 1 is reduced to 10 days, because the project duration will still be 25 days. The only way to reduce project duration (i.e., "schedule compression") without changing project scope is via the following methods:

- **Fast tracking**—Scheduling tasks on the critical path in parallel, whenever possible, without violating task dependency rules.
- **Crashing**—Adding extra resources to tasks on the critical path. However, assigning the same task to two resources at the same time makes it uncertain who is doing what, reducing accountability of resources and their productivity. If it is possible to split a task into two or more independent parallel tasks and assign them to two different resources, then crashing may be used effectively.
- **Extending working hours**—This method is used quite often, but caution must be exercised not to do it for extended periods of time, because staff may become burned out, reducing their productivity.

The schedule is dynamic—and so is the critical path. Even if the duration of critical path 5 is reduced to 20 days, the project duration will not be reduced to 20 days. As soon as the duration becomes less than 23 days, path 3 will become the new critical path, with a duration of 23 days. Any subsequent reduction in the duration of path 5 (which is no longer "critical") will not achieve any benefits in the project schedule and the project duration will remain at 23 days.

8

Issue Management (P4) and Configuration Management (P3)

Issue Management Planning (P4)

Purpose

Issues are triggered risks or risks that have occurred which may affect project goals if they are not resolved in a timely and effective manner. Issues are not the same as scope change. Whereas scope change requires the participation of the delivery team and usually involves design modifications, many issues may be resolved by administrative means. A few examples of issues are:

- Staff or resource problems, such as lack of the necessary skills or team performance
- Lack of cooperation or slow response from the client or management
- Requirements problems (these usually start as issues, but eventually scope change will be initiated in order to resolve requirements problems)
- Any triggered risk, such as a large increase in the cost of a resource

The purpose of issue management is the identification and management of issues that come up during all project frames, establishing actions to resolve them and minimizing their impact on the project. The issue management plan identifies resources responsible for each issue resolution task, resources for escalation when needed and the target dates for the issue resolution. The Issue Management Planning (P4) process is triggered every time one or more issues come up. In some cases, when an issue requires scope change, a new scope change request (SCR) is triggered. An SCR will be triggered even if the issue is the result of a budget shortfall or the project slips its schedule without changing the project scope.

Unlike the quality, risk and other project processes, there is no advanced issue planning. Issue planning and issue resolution are processes which are triggered by every new issue. Issue resolution and tracking are done in the Construction Frame.

The Issue Management Planning (P4) process is a set of activities required to manage issues as they occur. Issues are faced by every project. The issue management process ensures that:

- All issues that come up during the project life cycle are identified and recorded
- A resolution owner is identified and responsibility is assigned for resolving each issue by a specified deadline
- The impact of each issue on the project is analyzed
- Issue resolutions are planned, approved and managed
- Project stakeholders are notified of the status of issues
- Issue escalation and closure follow the identified process
- The expended costs for issue resolution are budgeted
- Issue resolution triggers a new SCR when needed

Issues may arise due to:

- Loss of committed resources or their unavailability
- Incorrect estimates or planning
- Ambiguous or incomplete requirements
- Changing needs and circumstances
- Project team performance
- Third parties or suppliers not meeting requirements, delivery time or quality standards
- Occurrence of the previously identified and accepted risks (see Chapter 6)
- Slow or no response from the client in providing answers or signing off on documents
- Other reasons, such as:
  - Natural disasters
  - Labor actions
  - Changes in currency value or a financial crisis

If resolving an issue necessitates a project scope change, the issue should be closed and the Scope Change Control (P7) process used instead. The status of all issues is reviewed during the weekly scheduled status meetings with the client and decisions are made about escalation if an issue cannot be resolved in a timely manner.

Cost of Issues

Issues may impact the schedule, the budget, and/or project scope and quality, which in and of itself will require opening a new SCR. This affects the project cost and duration even in those cases where the client is not charged for the budget increases. If an issue is due to loss of resources, incorrect estimates, poor planning, faulty design, etc., where the party responsible is the project manager or delivery organization, the client is not usually charged for any expenses incurred, depending on the contract type. Also, if the issue is the occurrence of a previously identified and accepted risk, for which the risk response costs have already been allocated and accounted for in the budget, there should be no additional charges, even if the party responsible is the client. If the issue is due to suppliers, which are selected by the delivery organization, not meeting commitments, then an attempt may be made to recovery costs from the supplier by legal means, provided there is a legal contract between the delivery organization and the supplier. Otherwise, the responsible party will be the one that approved using that particular supplier. Also, there is no additional charge for the project manager's efforts beyond what is already charged for managing the project. The client's time can never be charged.

Increased costs due to issues that cause delays in project implementation for which the client is responsible should be charged to the client organization. For example, during the course of a project, the client is expected to work with the project manager by representing the business at all times and supporting the project in many ways. Some typical client activities include:

- Answering questions
- Signing off on various project documents
- Representation at project-related meetings
- Performing user testing and other user tasks
- Resolving issues assigned to the client team members

All the above activities, as well as the expected response times for answering questions and sign-offs, must be clearly stated in the statement of work. For example, the statement of work may specify that the client must provide answers in writing to the project manager's questions within 48 hours or as mutually agreed. Similarly, it may state that the client's sign-off on the project deliverables or refusal to sign off with reasons documented must be done within four working days or as mutually agreed.

The project manager must allow the client ample time to review documents before approval is expected. However, if a document requires four signatures in sequence and obtaining each signature takes, say, a week, then it will take a month to get all the signatures together. What will the delivery team be doing during all that time? All delivery team members assigned to the project must charge their time against the project; otherwise they may be reassigned to another project and it may be a long time before the project gets new staff. This lag time must be planned for use in other scheduled activities whenever possible.

NOTE While waiting for a client's action, it may be possible to continue the planned delivery team activities after making certain assumptions, such as approval will be granted, the document will be signed off on, etc. The client must have consented to working temporarily without approval or sign-off. If the assumption turns out to be incorrect, the work produced by the delivery team during that time may not be usable. This will contribute to the project exceeding the budget and the schedule slipping. If the client did not object to the delivery team doing that work, the time spent will be charged to the client. However, if the client did not consent, then the time waiting for approval or sign-off is not chargeable to the client.

Unfortunately, often clients answer questions, sign off on documents and resolve issues assigned to them well past the due date. It is a good idea to send a friendly reminder ahead of the due date, but other priorities may still prevent the client from responding in a timely manner. The client should be informed of charges for that time using the Scope Change Control (P7) process, which provides a basis for cost or schedule changes and also informs stakeholders about those changes.

Sometimes the issue lies with the delivery organization rather than the client. Even though the client should not be charged in this case for the cost of the issue resolution, the client must be aware of the impact on the schedule due to the issue. Methods to shorten the schedule were described in Chapter 7, but they are not perfect, because they usually increase cost, risk or both.

Issue Management Planning (P4) consists of the following processes:

1. Identify/Modify and Document Issue (P4-1)
2. Assign Ownership (P4-2)
3. Propose or Modify Issue Resolution (P4-3)
4. Escalate Issue (P4-4)
5. Close Issue (P4-5)

Issue Management Process Flow

An issue may arise at any time. If resolution can wait until the project status meeting with the client, it should be brought up during that meeting. If the issue is urgent, a special meeting should be arranged.

When a new issue is triggered, the process flow enters the Identify/Modify and Document Issue (P4-1) process. The issue is discussed at the meeting and the project manager enters the initial record in the issue log form, a sample of which is provided in Table 8.1. The issue number, date, initial status of the issue (open) and a brief description are recorded. Issue status is indicated as follows: open (O), tracked (T), escalated (E), closed with no SCR opened (C) and closed due to opened SCR (C-SCR). Additional information required to understand the issue or contact information is entered in the comments column. Also, agreement must be reached on the due date by which issue solution is required. The issue log must be documented in the project control book and updated as the status of an issue changes or additional information about the issue is available.

Table 8.1 Issue Log Form

Issue Log						
Project Name:						
Issue #	**Date**	**Status**	**Brief Description**	**Issue Owner**	**Due Date**	**Comments**

The Issue Management Planning (P4) process flow, as shown in Figure 8.1, enters the Assign Ownership (P4-2) process, where it is acknowledged that the issue requires a solution and a qualified person from the delivery or client team is assigned responsibility for resolving it. The name of the issue owner is entered in the corresponding column of the issue log. If the issue owner is assigned on a different date than when the issue was first recorded, a second row must be added to the issue log with the same issue number, the new date and the issue owner's name.

Once the issue owner is assigned, he or she must propose a solution and due date for issue resolution in the Propose or Modify Issue Resolution (P4-3) process, and the flow is directed to a control point question (scope change required?). If it is obvious that an SCR will be required to resolve the issue, the answer is yes, and the issue will be closed in the Close Issue (P4-5) process and a new record is entered in the issue log with the status C-SCR. The output of process P4-5 will return the process flow back to the Construction Frame entry point, waiting for the next issue to track. At the same time, a control point question (issue resolved?) is asked. Since the issue cannot be resolved without an SCR, the answer is no, and a request will be issued to trigger a new SCR via exit point 15.

When it is not clear whether an SCR is required, then the answer to the control point question (scope change required?) is no. It will be possible to open an SCR later from the Construction Frame, so it is better to answer no if there is any doubt. In this case, the flow is directed to the Construction Frame for issue resolution and tracking via exit point 7F. The project manager changes the issue status to T in a new record, which must be entered in the issue log.

When tracking an issue in the Construction Frame, one of three options will be selected, and then recorded in the issue log and documented in the project control book:

- **Issue is resolved**—The notification comes from the Construction Frame via entry point 17. The project team and all affected groups must agree that the issue has indeed been resolved. If this is the case, the answer to the decision point question (issue resolution agreed?) is yes, and the issue will be closed with the status C in the Close Issue (P4-5) process. The process flow returns to the Construction

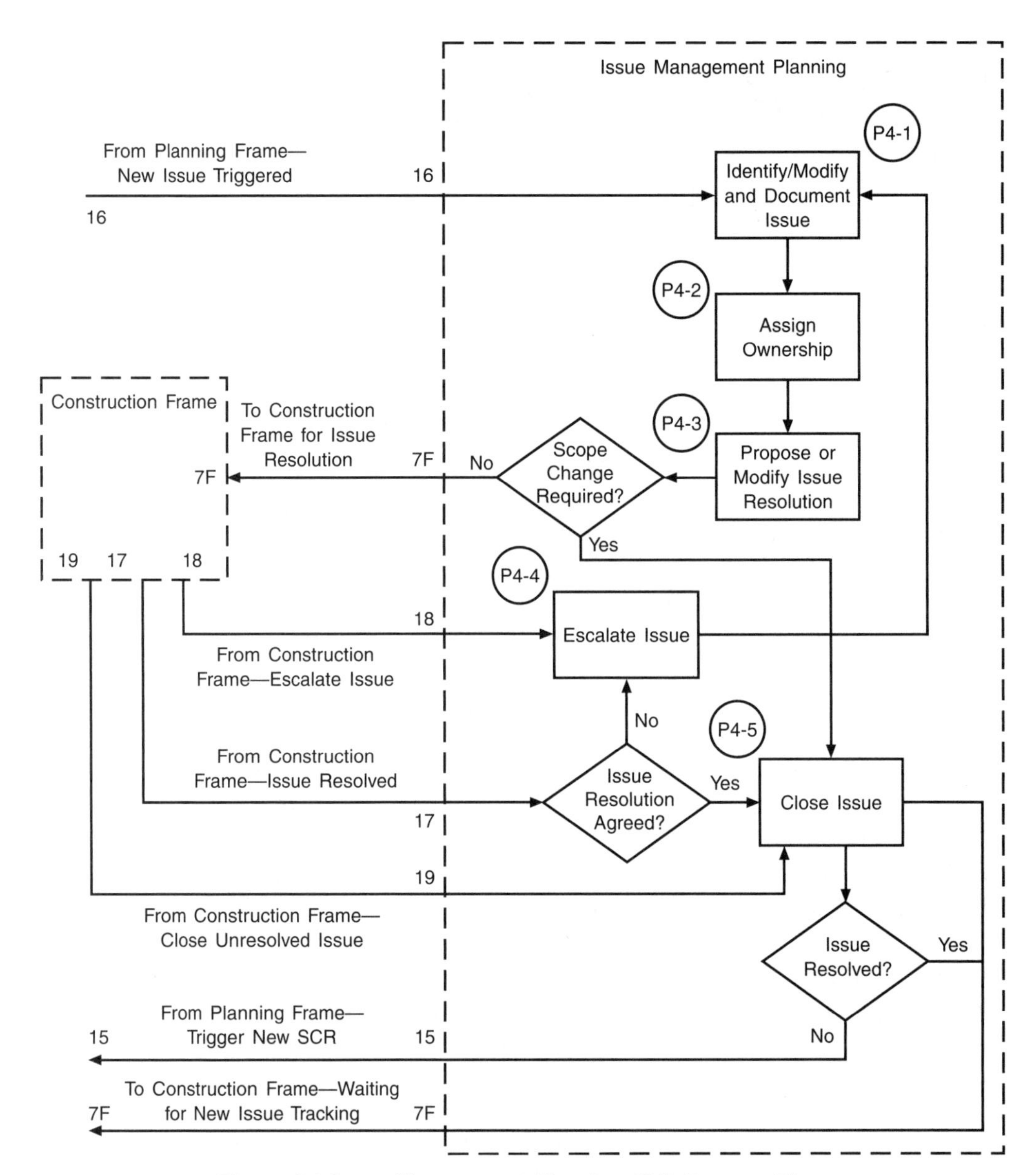

Figure 8.1 Issue Management Planning (P4) Process Flow

Frame entry point, waiting for the next issue to implement and track. If the answer to the decision point question (issue resolution agreed?) is no, the issue must be escalated in the Escalate Issue (P4-4) process. The issue log is updated with the new status E and stored in the project control book.

- **Issue can be resolved but some help from management is required or the issue owner is not able to deliver a solution on time**—In this case, the request to escalate the issue will come from the Construction Frame via entry point 18 to the Escalate Issue (P4-4) process. During this process, management is notified as described below and takes corrective measures. After the issue is escalated, some changes may occur, subject to escalation results. The issue may be modified, the issue owner may be changed (sometimes changing the issue owner helps resolve the issue) or the issue resolution proposal may be modified. Therefore, after escalation, the process flow returns to process P4-1 and the issue resolution process starts again.

- **Issue cannot be resolved by administrative actions**—This means that an SCR must be triggered to resolve the issue. This decision comes from the Construction Frame via entry point 19. The issue will be closed with the status C-SCR in process P4-5. The process flow returns to the Construction Frame, waiting for a new issue to track. At the same time, via a control point question (issue resolved?), the request will trigger a new SCR via exit point 15.

While the issue implementation is being tracked in the Construction Frame, the status of the issue is monitored and reported at least weekly. Issues are usually discussed during the regularly scheduled project status reviews. Those reviews are planned in accordance with the Communication Management Planning (P6) process and performed as a part of the Construction Frame process. Shortly before the issue resolution due date, the project manager should send a friendly reminder to the issue owner.

Escalation Considerations

Escalation should be avoided whenever possible. Corrective action should be taken before a situation gets to the point of requiring escalation. Good judgment must be used by the project manager to invoke escalation when and only when it is required. While escalation itself does not require stakeholder notification, stakeholders will be notified by the Scope Change Control (P7) process if the project cost or schedule is impacted and an SCR is issued. This will keep stakeholders informed about the situation, preventing unpleasant surprises at later stages of the project. When an issue is escalated to business management, the project manager must inform the client about this in a friendly and nonthreatening manner. However, if the project manager avoids escalating issues, those issues may remain unresolved and become his or her liability. An issue is escalated in accordance with the following considerations:

- If the issue is not resolved on time due to the action, inaction or inability of the issue owner, who is a client team member, the project manager will escalate the issue to the lead client using the issue escalation form in Table 8.2. If a decision about the course of action to resolve the issue is not made on time by the client, the project manager will present the situation to the delivery manager, who in turn will escalate the issue to the client's manager.
- If the issue is not resolved due to the action or inaction of a delivery organization member who does not report to the project manager, then the project manager will escalate the issue to the delivery manager, who in turn will escalate it to the resource owner or a senior manager.

Table 8.2 Issue Escalation Form

Issue Escalation	
Project Name: ______________________	Issue #: ______________________
Escalated to: ______________________ Escalated by: ______________________	Escalation Date: ______________
Issue Owner: ______________________ Date Raised: ______________________	Resolution Due Date: ___________
Brief Description:	
Impact on Project:	
Proposed Solution:	
Escalation Results:	
Comments:	

- If the issue is due to a supplier with which the delivery organization has a formal contract, then the project manager will escalate the issue to the delivery manager. The manager will bring the issue to the appropriate manager who is responsible for issuing the contract with the supplier, who will use the contract as leverage for escalation with the supplier.
- If the issue is due to the action or inaction of a delivery team member who reports to the project manager, then the project manager will escalate the issue to the functional manager in charge of the team member in question.
- If the issue is due to the project manager's error in planning, estimating, risk assessment, etc., no escalation takes place, but the issue must be reported to the delivery manager at the next status meeting.

Most often, the escalation process involves arbitration between the project manager and the party that caused the issue. As the result of escalation, the issue itself may change, the priority of team members' work may change, issue ownership may change or the issue resolution proposal may be modified. This will allow the issue to be resolved more expediently. In order to escalate an issue, an issue escalation form (see Table 8.2) must be completed.

Issue Management Forms

Issue management forms include the following fields:

1. **Issue #**—A sequential number assigned to each issue
2. **Project name**—Name of the project
3. **Escalated to**—Name of the person to whom the issue has been escalated
4. **Escalation date**—Date of escalation
5. **Issue owner**—Name of the person responsible for resolution of the issue
6. **Date raised**—Date when the issue was raised
7. **Resolution due date**—Date committed to for resolution of the issue
8. **Issue description**—Description of the issue
9. **Impact on project**—Impact on the project if the issue is not resolved
10. **Proposed solution**—Detailed description of the proposed solution
11. **Escalation results**—Results of the escalation
12. **Comments**—Any comments, including especially the context of the issue, deemed necessary to provide clarity on the issue

Issue Management Process Metrics

This procedure uses the following measurements (the cost of an issue does not include the cost of an SCR):

- Plan versus actual date for each issue resolution
- Total cumulative number of issues
- Number of issues converted into SCRs
- Cost of each issue broken down by the organization bearing the cost
- Impact of each issue on the schedule
- Cumulative cost of all issues
- Cumulative impact of all issues on the schedule

Configuration Management Planning (P3)

Purpose

The purpose of configuration management (CM) is to control project components at any given time and to prevent unauthorized changes to the completed project deliverables, processes or documentation. CM keeps track of and controls changes to these elements. It also maintains traceability and integrity of project elements. The Configuration Management Planning (P3) process produces the schedule of CM reviews.

CM is mandatory in software maintenance projects. Because many development platforms for software development projects have CM built into them, much of what is described in this section may not be applicable. CM also has limited use in electrical hardware, mechanical engineering, construction and other types of projects. In some cases, CM as described here may not be applicable at all.

Configuration Management Environment Structure

Some authors identify CM with change control in general and others with the project environment in particular. In this book, change control is viewed as a separate process and is described in detail in Chapter 10. Change control is used by CM when a scope change is authorized and the completed project deliverables have to be modified.

CM is indeed a specific part of the project environment, albeit limited in size. The environment is usually established before the beginning of a project, and a part of that environment is issued for CM purposes. For example, a simplistic explanation for software projects is there are development, test and completed module libraries (also called stage libraries). Developers use their individual development library areas while developing software modules and doing unit testing and then promote the modules to the test library in order to do an integrated test. They are not allowed to modify software in the test library. Rather, developers check out modules from the test library, modify them in the development library and after completion "promote" them again to the test library. When tests are completed, only a specifically authorized person is allowed to promote tested modules from the test to the stage library, which is a copy of the production library without using actual production data. This arrangement prevents anyone from moving modified modules into the stage library without authorization. When the project is complete, all modules are moved by the authorized person from the stage to the production library for business as usual operations.

Table 8.3 may be used to track this activity. Team member access authority to libraries is identified by a path to their location. The authority column lists the actions (create/modify/delete and promote to) that can be done with modules.

The development column shows the generic names (developer 1, developer 2, etc.) of staff members authorized to create/modify/delete and promote modules. In the promote to row in Table 8.3, N/A is indicated, which means there are no promotions to the development library. The create/modify/delete row shows a path to the individual

Table 8.3 Components Library Access Authority List

Authority	Development	Test	Stage	Production
	Individual Development Libraries	Test Library	Stage Library	Production Library
Create/Modify/Delete	Development Library 1: Developer 1; Development Library 2: Developer 2; Development Library 3: Developer 3	No	No	No
Promote to	N/A	Developer 1 Developer 2 Developer 3	Promoter A	Promoter B

development library and the names of developers who are authorized to use that library. Here, promoter A and promoter B are staff members with promotion authority to the stage and production libraries, respectively. Since usually there are multiple developers, there are also multiple libraries in the development area.

The test, stage and production columns show a path to the test, stage and production libraries and the names of team members who are authorized to create/modify/delete and promote modules to those libraries. No one is granted authority to create, modify or delete modules in any library except the development library. In the promote to row are the names of team members who are granted promotion authority to the test, stage or production libraries.

For tangible deliverables in projects other than software, instead of libraries, a special locked area of a warehouse is used for completed assembled products moved from the assembly and test areas.

Configuration Management Process Components

The Configuration Management Planning (P3) process is shown in Figure 8.2 and consists of the following components:

1. Establish Configuration Control Board (CCB) (P3-1)
2. Develop CM Plan (P3-2)
3. Establish CM Baseline (P3-3)
4. Conduct CM Baseline Audit (P3-4)
5. Approve Promotion (to stage/production libraries or warehouse) (P3-5)
6. Escalate (P3-6)
7. Correct Discrepancies (P3-7)

Establish Configuration Control Board (P3-1)

The function of the CCB is to approve the priority of required changes and their promotion to the stage library and production. For maintenance projects, the CCB usually consists of the project manager, the client representative, the quality assurance representative and also representatives of the stakeholders affected by the change. For development projects, promotion to the stage library is approved by the project manager or by an authorized person, while promotion to production is approved by the lead client. Approval authority must be documented in the project control book.

Develop CM Plan (P3-2)

For maintenance projects, the CM plan is a separate document, which prioritizes all required changes and provides a schedule for each new release to production. Usually, changes are released to production in *batches*, rather than in a continuous stream. Based on the plan, the project manager prepares a separate project implementation schedule for each planned batch of changes.

In development projects, the project manager schedules only CM baseline creation and baseline audits, which are added to the existing project schedule. In this case, the

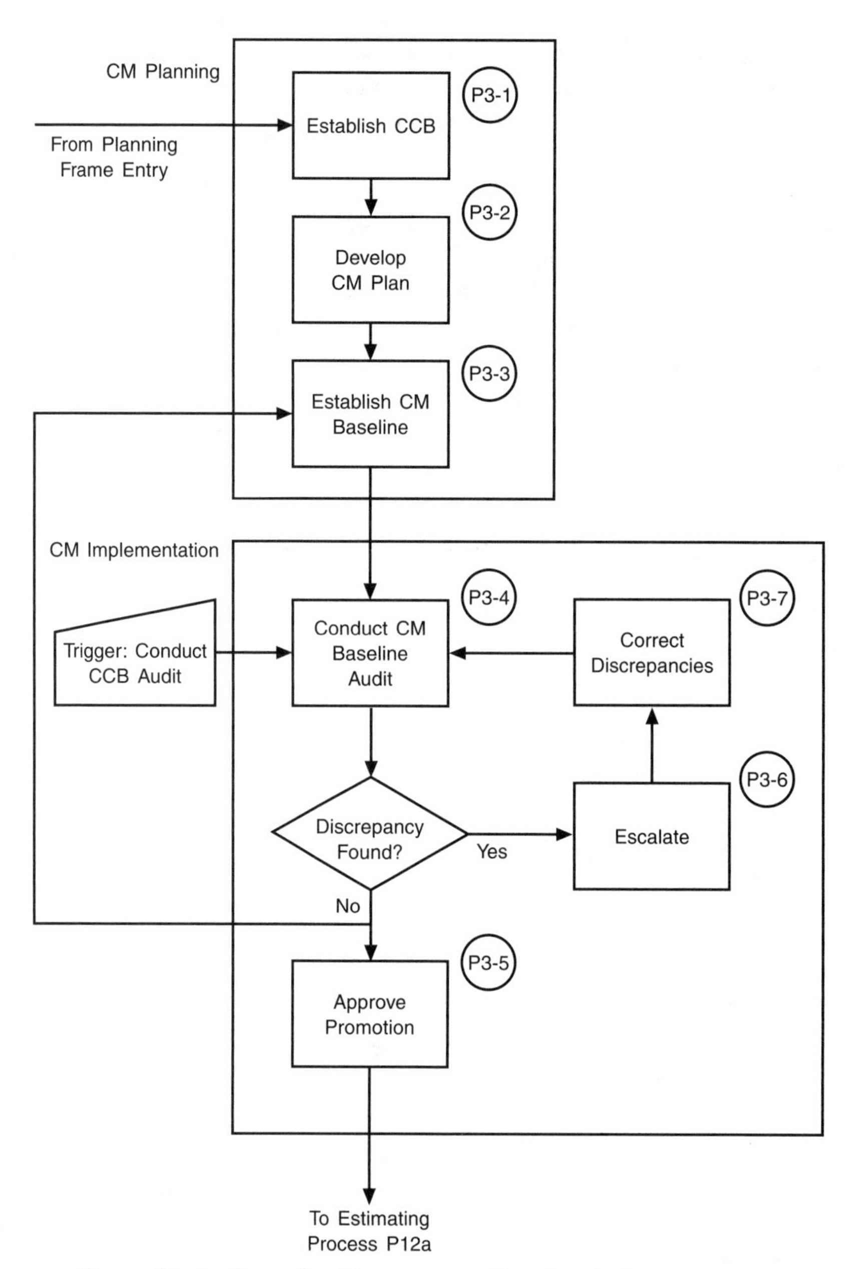

Figure 8.2 Configuration Management Planning (P3) Process Flow

CM plan does not have to be a separate document. The schedule of CM activities may be included in the overall project plan, and the CM processes and documentation should be recorded in the project control book.

In both types of projects, resources must be assigned to perform CM activities. The CM plan lists responsibilities for module modifications, test, promotions, approvals, baseline creation and audits.

Establish CM Baseline (P3-3)

The CM baseline is built differently for maintenance projects already in production and for current projects under development. For maintenance projects, it is the list of all project modules in the production library. A new CM baseline is established after each CM audit and documented in the project control book. For development projects, it is the list of completed modules added to the baseline as they are moved into the stage library or warehouse (for tangible deliverables) the first time. The list in the CM baseline contains the following information:

- Module name (name of program, database table, assembly, process, document, etc.)
- Module type (Java program, C++ program, database table, document, assembly, subassembly, etc.)
- Current version of module
- When last moved to production/stage/warehouse
- Location (library name, warehouse, etc.)

Conduct CM Baseline Audit (P3-4)

A CM audit is a comparison between the baseline and the current set of components. A CM audit must be conducted regularly. Any differences between the baseline and the current set of components must be documented in the promotion history. If any differences found are not documented, it indicates that unauthorized modifications were made to the completed modules or assemblies. If this is the case, the noncompliance must be escalated in process P3-6. The frequency of audits depends on the frequency of the deliverables updates. The audit should be done shortly before a scheduled production update or once every six months if there were no production updates. After a successful audit, if the answer to the control point question (discrepancy found?) is no, the flow returns to the process Establish CM Baseline (P3-3) and a new baseline will be created and documented.

Approve Promotion (P3-5)

Even though some team members are granted promotion authority to the stage and production libraries, no promotion to the stage or production libraries or placement in the warehouse is allowed without CCB approval. A memo must be sent to the authorized person with a request to promote specific modules from the test to the stage library or from the stage to the production library. For projects that are not software centric, approval must be granted to place the properly labeled and dated module in a warehouse. Corresponding approval must be granted and documented in the project control book.

Escalate (P3-6)

If the answer to the control point question (discrepancy found?) is yes, this means undocumented differences in modules were found between the baseline and the actual modules during the audit or the module counts do not match, and therefore the issue must be escalated. The discrepancy may be different time stamps or version numbers of a module between the last approval and the audit. Different time stamps may indicate that unauthorized changes were made to a module. Since libraries are dynamic and many promotions take place between two subsequent audits, all those promotions must be documented after their approval by the CCB, which makes all changes traceable. During escalation, the CCB, the project manager and the technical manager are notified of the discrepancy. The technical manager will assign a team member to conduct an investigation and find the reason for the noncompliance.

Correct Discrepancies (P3-7)

If undocumented changes to modules or unapproved promotions are found, corrective measures must be taken to reverse the unauthorized actions in the current scenario and to assure this will not happen again in the future. If changes were made in accordance with an approved SCR but no CCB approval was obtained, then CCB approval must be secured. Otherwise, all changes must be reversed. The CCB and the technical manager will be notified by the investigator. When problems are resolved, a new CM baseline audit must be conducted using process P3-4.

Configuration Management Metrics

The following metrics must be collected and documented in the project control book:

- CM plan versus actual effort and dates for all CM activities
- Number of promotions to the stage and production libraries
- Variations from the baseline and noncompliances detected during an audit

Configuration Management Documentation

The following CM documentation must be stored in the project control book:

- Names of and contact information for all CCB members
- CCB promotion approval documents
- Promotion history to the stage and production libraries
- CM plan and schedule of releases to client
- CM baseline
- CM baseline audit
- CM metrics

9

Communication Management (P6)

Purpose

The purpose of communication management is to define methods of communication between project team members of the delivery, outsourcing and business organizations in order to generate and exchange project-related information and to facilitate understanding between the sender of information and the receiver.

The Communication Management Planning (P6) process is a tool for the proper identification of stakeholders and developing project reporting and templates for different types of project communication, as well as scheduling effective communication to all stakeholders.

Communication Channels

Poor project communication greatly contributes to project failure, because it becomes impossible to resolve differences in expectations between the delivery team and stakeholders. The main reason for poor communication is insufficient time, when team members are busy and other priorities do not (apparently) allow communication in a timely manner. In fact, communication should be given high priority—it is often its absence which caused the conflicting issues to arise in the first place. Other contributors to poor communication include cultural differences, time zone challenges and in some cases the pure volume and intensity of the information that must be exchanged.

During the course of the project, the project manager spends much of his or her time communicating with the client, management, team members, suppliers, subject matter experts and so on. Delivery team members must communicate among themselves. Support personnel must communicate with team members and clients. Everybody may have to communicate with everybody else, but in large project teams it is almost impossible for the project manager to pay enough attention to everyone and their diverse needs for differing information at different times. Therefore, it is imperative to maximize the effectiveness of the communication channels that convey the most important project information.

The formula for calculating communication channels and links is

$$N * (N - 1)/2$$

where N is the number of people involved in communication. Thus, in a team of 5 members, the number of two-way communication channels is

$$5 * (5 - 1)/2 = 10$$

In a team of 10 there are 45 channels, and in a team of 23 there are 253 communication channels. It is obvious that the number of channels increases exponentially with the number of members. This may result in unmanageable communication among all team members. The technique often used to resolve this problem is to split large teams into manageable groups, with each team lead serving as a focal point for all communication outside that group. For example, suppose a team has a project manager, 3 team leads and 23 team members. If the team members are split into 3 groups, with 8 team members in the first and second groups and 7 members in the third group, then there are 6 communication channels between the project manager and 3 team leads, plus 28 * 2 + 21 communication channels in each team, for a total of 28 * 2 + 21 + 6 = 83 communication channels. This is significantly less than the 253 channels for the same number of team members all on one team.

Stakeholders

Stakeholders are members of the delivery, business, outsourcing or other organization who have a stake in the project or those whose interests may be positively or negatively affected by the project. We include here that time during which the project is under execution as well as the end product's steady-state or production operation. In fact, even your family may be stakeholders in your project, since their lives will be affected if you lose your job. The latest edition of the *PMBOK® Guide* even dedicates a new knowledge area to stakeholder management. As an example, if your project is the automation of a manual assembly line, which will reduce assembly personnel tenfold, then obviously many assembly line managers will strenuously object to the project and will do everything possible to delay or block it.

Table 9.1 Stakeholders List

#	Name	Position/ Organization	Stake in the Project	Phone	e-mail

Different stakeholders will have different interests in a project, sometimes mutually exclusive. This is why it is important to analyze project stakeholders—to understand their influence, interests, and attitude toward the project—and to develop plans to extract the greatest benefit from communication with them. Fully understanding them also may help to neutralize opposition to the project from important stakeholders.

You should try to sway powerful stakeholders to support your project right at the beginning, by highlighting how each one of them would benefit from the project. This may help you in getting required resources on time and gaining their support when necessary, which improves chances of your project's success. If you anticipate stakeholder reaction, you may use it in your communication, enhancing support and managing the opposition.

Project stakeholders should be identified and documented (see Table 9.1) in order of their power and influence. Stakeholders' motives, interests, influence and preferred method of communication must be analyzed, so as to enable the project manager to develop specific ways of communicating with them.

Communication Types

There are several types of project communication:

1. Proactive communication
2. Meetings and reports
3. Orientation and training

Proactive Communication

When people are unpleasantly surprised or receive bad news at the last minute, it can become a serious problem, damaging business relationships not only between the delivery team and the client, but also between different groups within the delivery organization. For example, if the resource manager informs the project manager at the last moment that the resources committed for project activities are not available, the project manager becomes frustrated and often does not have a chance to develop an alternative. The client will feel similarly if informed by the project manager at a regular status meeting that the project will be delayed and no new date can be set because the planned resources are not available. This has the dual negative effect of *surprising* and *alienating* key client stakeholders.

Proactive communication means that the affected side is informed as soon as a new risk is discovered, long before it becomes an issue. Both sides will have time to review alternatives and develop risk remediation plans.

The same is true if the client anticipates significant changes in the project scope. Learning about a scope change at the last moment may lead to getting stuck with unused resources or lack of the required resources, when there is no time to resolve the situation.

Poor communication also happens *within* the delivery team. Delivery team members may be confused about details of their assignment due to insufficient communication between the project manager and the team members. The delivery team members must proactively communicate the lack of information before errors are made. It is the project manager's job to set these communication expectations early in the project.

Proactive communication is often informal communication, such as a phone call, business lunch or even a short chat in the hall. If a topic of discussion is important enough, it may be necessary to send an e-mail to the source of the information with a brief outline of the subject discussed and decisions made. This approach should be used with caution, because it may be counterproductive to document the subject of a confidential conversation, especially with influential stakeholders.

Recommendations for what types of media are best for various project communications can be found in *A Project Manager's Guide to Emotional Intelligence* by Anthony Mersino.

A communication plan should include a simple guide or matrix which shows the preferred types of media and periodicity of communication for key project information exchange, identifying the intent and audience for each exchange of information.

Meetings and Reports

Meetings take place when several people are required to get together in order to make a common decision, conduct a review or discuss an issue. Meetings may be scheduled in advance in accordance with the project plan or called on short notice due to the occurrence of an urgent event. Status meetings should always be accompanied by status reports.

Planned Meetings and Reports

Planned meetings are meetings that are planned in advance and appear in the project plan. Those meetings are associated with planned events in the project, like the end of a project frame, the completion of a deliverable, etc. Some planned meetings take place on specific dates and are scheduled long in advance. Many types of planned meetings are described in detail in other sections of this book, such as quality control reviews, quality assurance audits, business requirements document review and others. Other planned meetings are several types of the project status meetings. The process that must be followed in all planned meetings is as follows:

1. Send meeting notification and invite participants (include all logistics, such as the venue, start time, end time, any required preparation, and any multimedia links)
2. Send meeting agenda and applicable documentation or reports
3. Conduct meeting and take meeting minutes
4. Distribute meeting minutes
5. Receive corrections to recorded minutes
6. Document the meeting in the project control book
7. Take action according to meeting decisions

The five types of planned status review meetings and their frequency for the duration of a project are:

1. Delivery team status review: weekly
2. Client status review: weekly
3. Business executive status review: monthly
4. Delivery executive status review: monthly
5. Outsourcing/offshore team status review: weekly

Status reviews may be face-to-face meetings, teleconferences, videoconferences or a combination. Consideration must be given to the economic and ecological attributes that each of these meeting types has to offer.

Delivery Team Status Review Meeting and Report

The objectives of the delivery team status review meeting are to review the progress of each delivery team member and overall project progress for the past week. Each team member will report on the following items:

- Weekly status
- Active action items assigned
- Progress on assigned tasks and task actuals versus plan
- Open issues and progress in their resolution
- New risks and progress in risk remediation planning
- Quality control results
- Plan for the next two weeks and the forecast for its implementation

After individual review is completed, the following subjects will be discussed with the entire team:

- Overall project status against the plan
- Overall project risks and issues
- Active change requests
- Status of project quality assurance audits
- Status of quality control reviews
- New and unresolved team conflicts (it is important to handle conflict carefully; some conflicts must be handled on a face-to-face basis and not publicly)

- Overall management decisions related to the project
- Client satisfaction and feedback

If the delivery team is split into several subteams, each with a separate team lead, then the individual team member review will be conducted by the team lead and the overall team review by the project manager. Team leads will provide a summary of the individual team member reviews, highlighting the most essential information.

The delivery team status review meeting must be scheduled in advance to run weekly as a face-to-face meeting or teleconference. If the review is held via teleconference, then status review reports must be received by the project manager or team lead(s) at least several hours in advance. At the start of the meeting, a scribe must be assigned to take notes.

The delivery team status report (see Table 9.2), completed by each delivery team member, must be filed in the project control book as a spreadsheet in electronic form and a hard copy must be available for a face-to-face meeting. This status report consists of two parts:

1. Work performed by the team member for the reporting period
2. Work forecast for the next two weeks

The following are the fields for the first part:

1. Task ID
2. Scheduled task name for the week reported
3. Percent completed (with a metric to represent this, which has been agreed upon up front)
4. Scheduled start date
5. Actual start date
6. Scheduled end date
7. Planned/actual end date
8. Budgeted hours
9. Actual hours worked so far
10. Additional time needed (hours)
11. Reason for delay (if any)
12. Comments/issues/concerns requiring management attention

The following are the fields for the second part of the report (next two weeks):

1. Task ID
2. Planned activities/tasks
3. Scheduled start date
4. Actual start date (if started)
5. Scheduled end date
6. Budgeted hours
7. Comments

Table 9.2 Delivery Team Status Report Form

Delivery Team Status Report

Project Name: ____________________

Weekly Status Report for Week Ending on: ____________________

Name: ____________________

Task ID	Scheduled Task Name for the Week Reported	% Complete	Sched. Start Date	Actual Start Date	Sched. End Date	Planned/ Actual End Date	Budgeted Hours	Actual Hours Worked	Add'l Time Needed (Hours)	Reason for Delay

Comments/Issues/Concerns
Requiring Management Attention:

Task ID	Planned Activities/Tasks for the Next Two Weeks	Scheduled Start Date	Actual Start Date	Scheduled End Date	Budgeted Hours	Comments

Client Status Review Meeting and Report

The purpose of the client status review meeting is to provide the client with an update on the project status and to discuss all open issues, change requests and new risks. Also, client satisfaction will be discussed at least once a month.

It is preferable to update the client on the status of project milestones rather than to tediously review detailed tasks that do not have business meaning to the client. This is done in order to avoid client attempts to micromanage the project and having to explain technical issues to a nontechnical client. At the end of the review, the client should have good visibility of the project and the project issues. It is worthwhile to verify that the client agrees that the needed visibility is being provided.

The client status review meeting must be scheduled in advance to run weekly either as a face-to-face meeting or teleconference. If the review is held via teleconference, then the status review report and related documentation must be received by the client at least several hours in advance. At the start of the meeting, a scribe must be assigned to take notes.

The client satisfaction discussion will be held monthly and is based on results provided by the client satisfaction review tool, as discussed in Chapter 14.

A template for the client status report is provided in Table 9.3.

Business Executive Status Review Meeting and Report

The business executive status review is the same as the client status review meeting, but with the business executive present. Since the meeting is held once a month, the reporting period for the report is monthly. The meeting may be combined with the client status review meeting, in which case only the business executive status review report is provided. This meeting is required to ensure that the project is aligned with the corporate business strategy and to keep the project properly visible to business executives. Since a business executive is an influential business stakeholder, it is wise to know his or her project vision and maintain the executive's support.

The project manager must be realistic about the business executive attending the review once a month, since the executive will attend only meetings of high priority. Attendance depends on the visibility of the project, its size and the effect of the project on the business. Nevertheless, a business executive must be invited to participate well in advance, and the meeting agenda and materials must be sent out at least two days in advance.

The business executive status report is the same as the client status report, except that the reporting period field must indicate one month, rather than one week as in the client status report.

Delivery Executive Status Review Meeting and Report

The objective of the delivery executive status review meeting is to keep delivery executives up to date about the project health. This monthly meeting should be scheduled by the executive office, which invites between five to eight project managers to report on their projects. These projects may be from a variety of business areas and may be un-

Table 9.3 Client Status Report Form

Client Status Report	
Project Summary	
Project Name: ______ Project Manager: ______ Client Name: ______ Reporting Period: ______ Planned Completion Date: ______ Total # of Scope Changes: ______ Total Cost of Scope Changes: ______	
Active Issues	
Brief Description of Issue	**Issue Status**
Major Milestone 1	
Milestone Name: ______ Planned Start Date: ______ Actual Start Date: ______ Planned End Date: ______ Actual End Date: ______ Status: ☐ Not Started ☐ In Process ☐ Completed Comments:	
Major Milestone 2	
Milestone Name: ______ Planned Start Date: ______ Actual Start Date: ______ Planned End Date: ______ Actual End Date: ______ Status: ☐ Not Started ☐ In Process ☐ Completed Comments:	
Major Milestone 3	
Milestone Name: ______ Planned Start Date: ______ Actual Start Date: ______ Planned End Date: ______ Actual End Date: ______ Status: ☐ Not Started ☐ In Process ☐ Completed Comments:	
Financial Summary	
Planned Project Cost: ______ Cost of All Scope Changes: ______ Actual Cost to Date: ______ Cost Outlook: ______ Comments:	
Client Satisfaction Review	
Date of the Last Review: ______ Satisfaction Rating: ______ Progress Made: ______ Comments:	

related. Each project manager must present one project in about five to ten minutes, walking through the delivery executive report (see Table 9.4) and answering questions, if any. The following information is presented at the meeting:

- High-level project perspective
- Major unresolved issues
- Major project risks
- Project quality audits
- Project financial health

Along with the executive and project managers, intermediate managers associated with the projects reported on also are in attendance.

A template for the delivery executive status review report is provided in Table 9.4.

Financial Status Report

The financial status of the project for the reporting period may be presented using the earned value analysis tool discussed in Chapter 15. It shows variances, performance indices, the projected cost of the project and the projected completion date, assuming that no significant performance changes will occur in the remaining time.

In order to see the dynamics of project performance for the preceding 12 months in graphical format, statistical data for planned value (PV), earned value (EV) and actual cost (AC) and the starting month of the statistical report are entered, as shown in Figure 9.1. PV, EV and AC values are taken from the earned value analysis tool.

Outsourcing/Offshore Team Status Review Meeting and Report

The objectives of the outsourcing/offshore team status review meeting are to control subproject implementation and ensure there will not be any negative impact on project quality, cost and schedule.

In your relations with an outsourcing/subcontractor team, you must see yourself as the client and the subcontractor team as the delivery team. Even though your team may be as technically capable as the subcontractor team, attempts to micromanage their project must be avoided. You probably will never see the team or talk to most of the team members. If the outsourcing/subcontractor is an offshore organization, you may not even see the project manager. Instead, you will meet weekly with the local representative in your physical area, who is usually a senior-level expert. Today, most offshore companies are located in India, China and Eastern Europe, and the time difference between your location and the offshore team may be between 7 and 14 hours. This makes it very difficult to hold regular teleconferences with the project manager, let alone team members, and reduces the subcontractor's visibility to the delivery team and the client.

In this environment it is of the utmost importance to provide the subcontractor with very detailed requirements so as to eliminate the slightest possibility of misunderstanding, because cultural preference may disallow—or at least discourage—the offshore team from asking for clarification. Experience with offshore teams clearly indicates that most

Table 9.4 Delivery Executive Status Review Report Form

Delivery Executive Status Review Report

Reporting Period: From ___/___/_____ (MM/DD/YY) to ___/___/_____ (MM/DD/YY)
Project Name: ______________________ Project Manager: ______________________
Client Business Area: ______________________ Client Name: ______________________
Frame: ________ Frame Budget: __________ Projected Cost of the Frame: ________

Current Frame Major Milestones

#	Milestone Name	Due Date	Done?	Actual Delivery Date	Schedule Performance Index	Cost Performance Index

Comments:

Financial Summary

Current Month Plan: ______________________ Actual Month Plan: ______________________

Current Frame Plan: ______________________ Actual Frame Plan: ______________________
Projected Current Frame Outlook: ______________________

Total Project Plan: ______________________ Actual Project Plan: ______________________
Projected Total Project Outlook: ______________________

Comments:

Failed Quality Assurance Audits in the Reporting Period

Audit Date: ______________________ Noncompliance Area: ______________________

Audit Date: ______________________ Noncompliance Area: ______________________

Comments and Root Cause Analysis of Noncompliance:

Corrective Action Taken:

Failed Quality Control Reviews in the Reporting Period

Deliverable Name: ____

Deliverable Name: ____

Comments and Root Cause Analysis of Failure:

Corrective Action Taken:

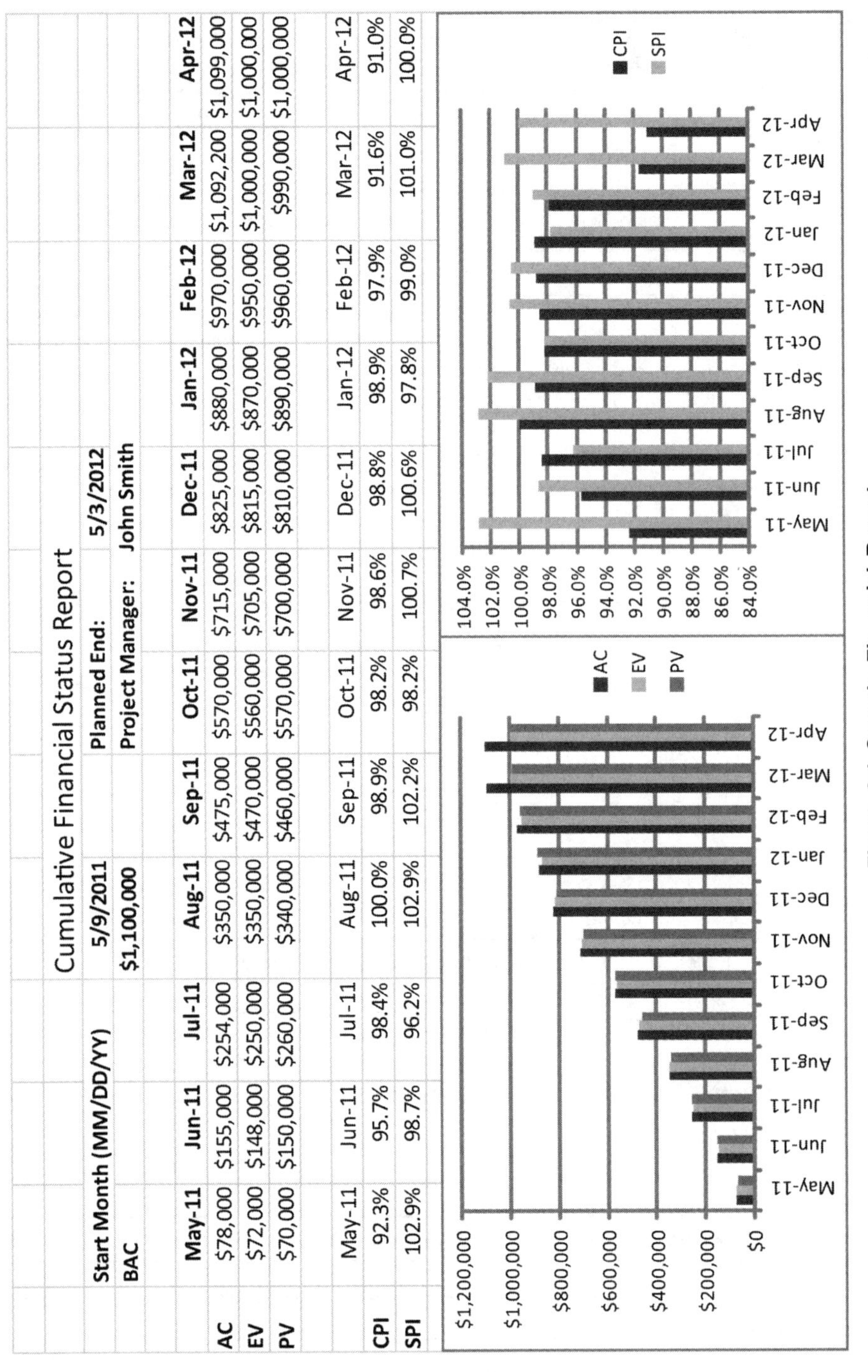

Cumulative Financial Status Report

Start Month (MM/DD/YY)	5/9/2011	Planned End:	5/3/2012
BAC	$1,100,000	Project Manager:	John Smith

	May-11	Jun-11	Jul-11	Aug-11	Sep-11	Oct-11	Nov-11	Dec-11	Jan-12	Feb-12	Mar-12	Apr-12
AC	$78,000	$155,000	$254,000	$350,000	$475,000	$570,000	$715,000	$825,000	$880,000	$970,000	$1,092,200	$1,099,000
EV	$72,000	$148,000	$250,000	$350,000	$470,000	$560,000	$705,000	$815,000	$870,000	$950,000	$1,000,000	$1,000,000
PV	$70,000	$150,000	$260,000	$340,000	$460,000	$570,000	$700,000	$810,000	$890,000	$960,000	$990,000	$1,000,000

	May-11	Jun-11	Jul-11	Aug-11	Sep-11	Oct-11	Nov-11	Dec-11	Jan-12	Feb-12	Mar-12	Apr-12
CPI	92.3%	95.7%	98.4%	100.0%	98.9%	98.2%	98.6%	98.8%	98.9%	97.9%	91.6%	91.0%
SPI	102.9%	98.7%	96.2%	102.9%	102.2%	98.2%	100.7%	100.6%	97.8%	99.0%	101.0%	100.0%

Figure 9.1 Sample Financial Report

problems with offshore deliverables are caused by poorly defined or misunderstood requirements. This can be avoided with more effective communication which considers cultural differences. The issue of cultural differences is explored in the book *Kiss, Bow or Shake Hands: The Guide to Doing Business in More than 60 Countries* by Terri Morrison and Wayne A. Conaway.

The following information will be reported at the weekly meeting with the outsourcing/offshore representatives:

- Weekly status
- Plan for the next two weeks
- Issues, risks and action items
- Performance and quality

A template for the outsourcing status review form is provided in Table 9.5.

Event-Driven Meetings

Event-driven meetings are meetings that have not been previously scheduled. They take place when the urgent resolution of an issue is required. Sometimes the need for an event-driven meeting arises during other meetings or even casual conversation or an e-mail exchange. The standard meeting process is followed:

1. Send meeting notification and invite participants (specify venue, purpose, agenda and logistics)
2. Send meeting agenda and applicable documentation or reports
3. Conduct meeting and take meeting minutes
4. Distribute meeting minutes
5. Receive corrections to recorded minutes
6. Document the meeting in the project control book
7. Take action according to meeting decisions

Sometimes the meeting must take place within hours after the issue was brought up, and there may not be enough time to send formal invitations and a detailed agenda. In this case, an impromptu conference call may be satisfactory, and only the most essential people should be brought in after rearranging priorities of other scheduled activities.

Orientation and Training

Orientation is required to provide both the process and project information to the project manager and the delivery team members, thereby introducing them to the project and instructing them about existing processes and/or the technology used in the project. Training is required to:

- Enable users of the project to use—and to help produce—project deliverables in the correct and proper way

Table 9.5 Outsourcing Status Review Report Form

Outsourcing Status Review Report	
Subproject Summary	
Subproject Name: ______________________	
Subcontractor: ______________	Project Manager: ______________
Reporting Period: ______________	Planned Completion Date: ______________
Total # of Scope Changes: __________	Total Cost of Scope Changes: __________
Active Issues	
Brief Description of Issue	**Issue Status**
Major Planned Subproject Milestone 1	
Milestone Name: ______________________	
Planned Start Date: ______________	Actual Start Date: ______________
Planned End Date: ______________	Actual End Date: ______________
Budgeted Hours: ______________	Actual Hours: ______________
Additional Hours Needed: __________	
Status: ☐ Not Started ☐ In Process ☐ Completed	
Comments:	
Major Planned Subproject Milestone 2	
Milestone Name: ______________________	
Planned Start Date: ______________	Actual Start Date: ______________
Planned End Date: ______________	Actual End Date: ______________
Budgeted Hours: ______________	Actual Hours: ______________
Additional Hours Needed: __________	
Status: ☐ Not Started ☐ In Process ☐ Completed	
Comments:	
Major Planned Subproject Milestone 3	
Milestone Name: ______________________	
Planned Start Date: ______________	Actual Start Date: ______________
Planned End Date: ______________	Actual End Date: ______________
Budgeted Hours: ______________	Actual Hours: ______________
Additional Hours Needed: __________	
Status: ☐ Not Started ☐ In Process ☐ Completed	
Comments:	
Financial Summary	
Planned Project Charges: __________	Charges for All Scope Changes: __________
Actual Charges to Date: __________	Project Outlook: ______________
Comments:	

- Assist the project manager and delivery team members to perform quality management activities
- Familiarize the project manager, delivery team members and client with some essential processes and tools to perform their duties, such as the scope change request process, client satisfaction review tool, risk assessment tool and others

Training materials must be prepared by qualified personnel. Training should be scheduled to include all required participants. For example, quality management training is performed by a member of the quality management team. Sometimes delivery team members need professional training, but this is rarely done within the project framework. If a budget for such training has not been provided, then this kind of training is not a part of the project.

The project manager's orientation usually takes place at the time of his or her assignment to the project. It may be given by the client, and sometimes a delivery manager may provide orientation in specific technical issues about which the client does not know much. One way to organize training is for the client to give a presentation, which must be documented in the project control book.

Delivery team orientation includes the following areas:

- Project standards, methodology and processes
- Tools used in the project
- Technical issues

The kickoff meeting, which is described in Chapter 15, is a special type of project orientation, which concentrates on the following:

- Project objectives, approach, schedule and facilities
- Roles and responsibilities

Meeting Agenda Form

Table 9.6 may be used as a meeting agenda form or as a checklist to ensure that all proper logistics are covered.

Meeting Minutes Form

A template for a meeting minutes form is provided in Table 9.7.

Communication Plan

The purpose of a communication plan is to establish specific rules for communication management in order to allow the exchange of project information among the project manager, delivery team, client, business users, management and other stakeholders.

Table 9.6 Meeting Agenda Form

Meeting Agenda	
Project Name:	
Meeting Initiator's Name:	
Date:	Time: ☐ AM ☐ PM
Location:	
Reason for the Meeting:	
List of Attendees:	
Agenda Item 1:	
Agenda Item 2:	

Table 9.7 Meeting Minutes Form

Meeting Minutes	
Project Name:	
Meeting Initiator's Name:	
Date:	Time: ☐ AM ☐ PM
Location:	
Reason for the Meeting:	
List of Attendees:	
Minutes for Agenda Item 1:	
Minutes for Agenda Item 2:	

There is no need to create a separate communication plan document as long as the following elements are documented in the project control book as they occur and their schedules are included in the project plan:

- Establish all project status meetings, including dates and periodicity, and specify the intent and audience of each
- Describe the content of status reports and the forms used to produce them
- Establish orientation and training schedule
- Develop and produce all training materials

10

Scope Change Control (P7)

Purpose

The project scope, as opposed to the product scope, is the work required to deliver the solution to the client. The product scope is the scope of the business solution, described in the business requirements document. Any change in business requirements is a product scope change, which will lead to a change in the required project work. The term project scope change usually is used for business requirements changes and is the subject of this chapter.

The purpose of the Scope Change Control (P7) process is to manage changes to business requirements and design and their effects on the project. Process P7 is also used when changes are required in the project cost and schedule due to poor project performance. The Scope Change Control (P7) process ensures that:

- Changes are identified, analyzed, approved, planned and implemented
- Project stakeholders are fully aware of all changes
- All scope change requests are documented and the approved changes are entered in the project plan

The Scope Change Control (P7) process is activated when a request to issue a new scope change is initiated by any of the following:

- Client team member
- Delivery team member

- Any stakeholder
- Issue management process
- Legally mandated requests

In order to analyze, plan and implement a scope change, the Scope Change Control (P7) process interacts with other processes in all project frames.

Scope Change Control (P7) Process

The Scope Change Control (P7) process describes the interaction details between the delivery and client teams at the time when a project scope change is required.

All too often, clients directly ask the delivery team members to implement scope changes, rather than following the scope change process. This is especially true, but no less insidious, when the scope changes are small. If team members accept these changes for implementation, the consequences of such requests, also called collectively "scope creep," can be very grave and may cause one or more of the following problems:

- Undocumented changes that could have significant technical, business, safety, environmental, social and/or legal implications.
- The scope change is implemented without a scope change requirements analysis.
- The cost of implementing the change is not covered by the existing budget.
- Implementation of the change is not incorporated in the project plan. Due to project dependencies, many other project activities may slip in the schedule or require extra work as a consequence of the change. Even if the change takes only a few hours to implement, many other project tasks may be delayed by hours each. This delay can easily cascade and multiply several times in the overall effort to incorporate the change, causing significant overall project slippage. The later in the project cycle the change is requested, the greater the cost and the greater the overall impact on the project.
- If the impact of the scope change on other tasks and projects is not thoroughly investigated, it may affect not only the project but also the steady-state operation of the organization.

Therefore, the project scope change process must be strictly followed by both the client and the delivery team members. This must be made very clear to everyone. Rules for enforcement of the Scope Change Control (P7) process must be included in the statement of work. Also, the delivery team members must be specifically instructed not to accept new change requests or modifications from anyone except the project manager or specifically authorized personnel.

Some clients and delivery team members may find it difficult to follow the change request process flow chart; therefore, the following description attempts to resolve this potential issue. This description should be included in the statement of work. The process is comprised of the following steps:

1. The scope change must be identified, documented and submitted to the project manager for approval using the scope change request (SCR) form. The SCR must identify all proposed changes to the project scope. If the scope change is identified by a business user or business manager, it must be reviewed by the lead client before submitting the SCR to the project manager. If the scope change is identified by a delivery team member or the delivery manager, the team lead will review the proposed scope change and make a decision whether to proceed with it or reject it. If the decision is to proceed, then the SCR is submitted to the project manager. An SCR will be also initiated and submitted when a modification to the project scope is proposed as part of a cost reduction or schedule issue in order to fit into the existing budget or schedule.
2. The SCR is reviewed by the project manager with the relevant delivery team members, and the effort required to analyze the SCR is estimated. The SCR may be accepted for further evaluation, deferred to a later date or rejected. The SCR must be documented in the project control book.
3. If the SCR is accepted, project stakeholders will be notified by the project manager. If the SCR is rejected, it is formally closed, with reasoning behind the decision provided. If the SCR is deferred, no action is taken until the deferral date. On the deferral date, the SCR is again sent to the project manager for approval and step 2 is repeated. The SCR status is documented in the project control book.
4. If the SCR is accepted, the budget for the SCR analysis is submitted to the client for approval. If the scope change is small (the estimate of effort for the SCR analysis is under two hours), budget approval for the SCR analysis is not required and step 5 is skipped.
5. If the budget for the SCR analysis is approved by the client, the SCR status will be documented in the project control book and the SCR analysis will be performed. If the budget is rejected, the reason for rejection must be documented and the SCR is closed.
6. The SCR planning and implementation tasks are estimated, documented in the project control book, and the budget is submitted to the client for approval.
7. If the SCR planning and implementation budget is approved, it is documented in the project control book and the SCR is incorporated in the overall project plan. If the budget is rejected, the reason for rejection must be documented and the SCR is closed.
8. The project plan is updated to include the SCR in the schedule, and the SCR is implemented in accordance with the project plan.
9. The completed SCR is closed. If serious issues or risks are discovered during planning or implementation, which jeopardize the SCR implementation, a new issue must be opened in order to mitigate the problem.

The only exception to this process is when the project falls behind schedule due to client action or inaction, such as a delay in signing off on project documentation or deliverables. In this case, no client approval is required, and the SCR is sent to the client for information only. The client will have the opportunity to escalate the issue with delivery management. At the same time, project stakeholders will be notified by the project manager.

NOTE When a new SCR is added to the schedule, some organizations allow project managers to baseline the schedule. The baseline is a reference point, which is compared with the project actuals to determine project health and perform earned value analysis. A project is usually baselined first after the project plan and the budget are approved and then again after the new, more detailed estimates have been made. From that point, variances in the cost and schedule can be tracked by comparing them to a baseline target. A new baseline is not established after each change request, because all old project tracking information would be erased. For example, if a project was in a disastrous state, establishing a new baseline would show it to be in perfect health with no variances in cost and schedule. Establishing a new baseline is allowed by senior management in exceptional cases. It is often allowed if too many change requests have been implemented and the Gantt chart no longer makes any sense.

Earned value analysis must be performed before entering new SCR information into the schedule and then again after scheduling a change request but before the beginning of its implementation. Before the second calculation, the budget at completion should be increased by the cost of the change request and the planned end date extended by the scheduled time needed for the SCR implementation. The results of both earned value analyses must be identical. All following earned value analyses will correctly reflect the health of the project.

The schedule, however, will show all tasks that were moved due to the SCR tasks as slipped—and the Gantt chart does not look pretty. To alleviate this, some project managers create a personal copy of the schedule and baseline it. The Gantt chart now may look pretty, but the slippage cost or schedule prior to the new baseline will no longer be displayed. This personal copy of the schedule is just for reference and should not be shared with anyone.

Scope Change Control (P7) Process Flow

The process flow moves across the Requirements, Planning and Construction frames. Scope Change Control (P7), as shown in Figure 10.1, includes the following processes:

1. Submit SCR (P7-1)
2. Review SCR for Evaluation (P7-2)
3. Notify Project Stakeholders (P7-3)
4. Approve Budget for SCR Analysis (P7-4)
5. Approve SCR Implementation (P7-5)
6. Close SCR (P7-6)
7. Wait for End of SCR Deferral (P7-7)

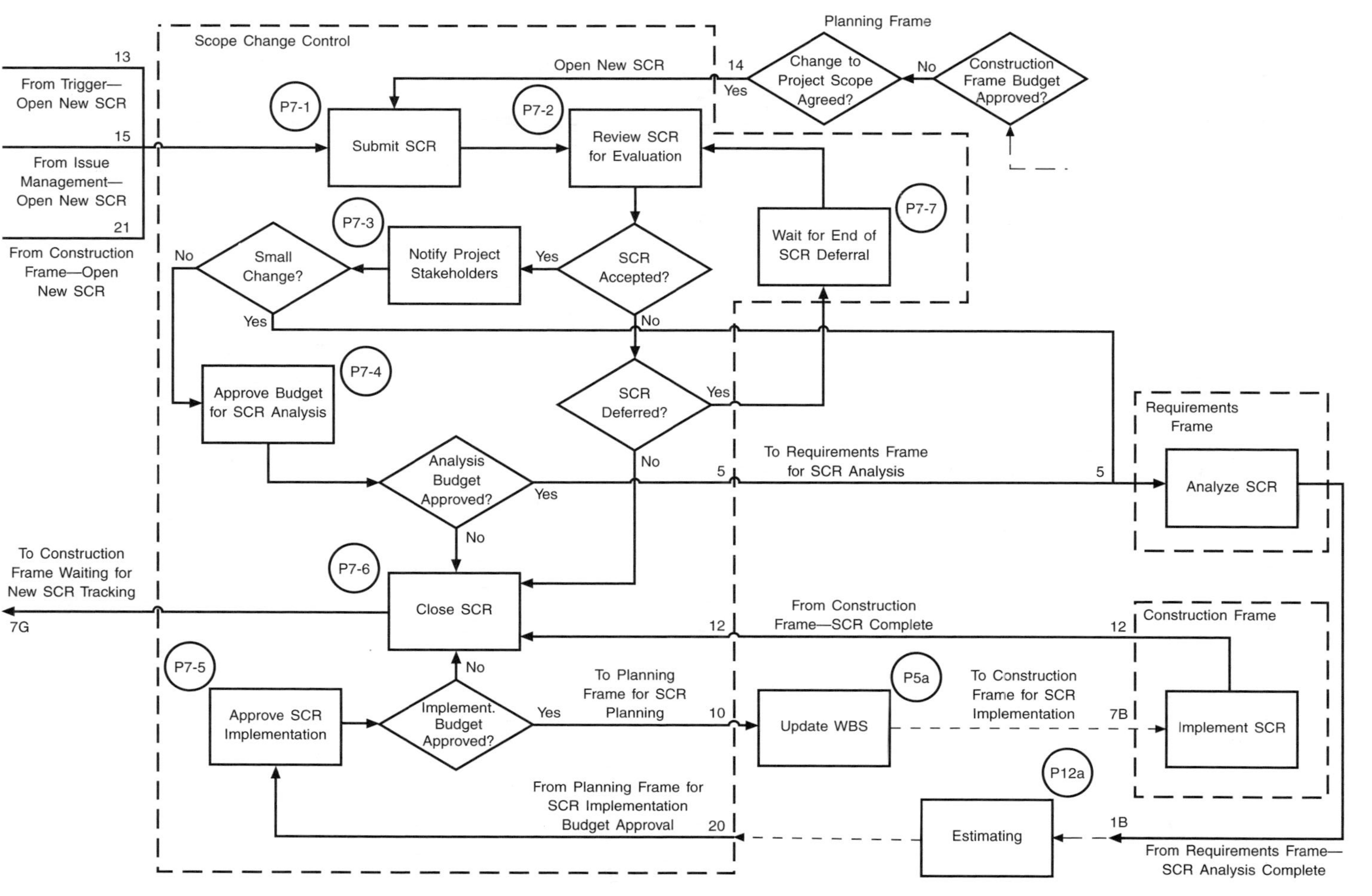

Figure 10.1 Scope Change Control (P7) Process Flow

In addition to the above seven processes, there are several additional processes that are logically a part of the Scope Change Control (P7) process but physically located outside of this process and even outside of the Planning Frame. Some of those additional processes are generic processes, such as estimating, build/update the work breakdown structure, etc., which are also used in the scope change planning flows. The additional processes used for scope change control are:

- Estimate SCR: Planning Frame
- Update work breakdown structure: Planning Frame
- Analyze SCR: Requirements Frame
- Implement SCR: Construction Frame

The Scope Change Control (P7) process is triggered via entry point 13 when a business user or delivery team member initiates a new SCR. If the initiator's team lead finds the SCR necessary, the Submit SCR (P7-1) process will be executed. Otherwise, no SCR is initiated and no further action is taken. Process P7-1 will also be executed in the following cases:

1. The request for a scope change comes from the Planning Frame via entry point 14, when the budget is not approved but agreement is reached to reduce the project scope in order to fit the project into the available budget and schedule.
2. The request for a scope change comes from the Issue Management Planning (P4) process via entry point 15, when the issue cannot be resolved without changing the project scope.
3. The request for a scope change comes from the Construction Frame or the Closing Frame via entry point 21, when poor project performance identified during tracking in the Construction Frame cannot be improved without a scope change and also if a defect found during testing cannot be fixed without a scope change.

Once the SCR is submitted, the Review SCR for Evaluation (P7-2) process is executed. The SCR is reviewed with the delivery team members and is either accepted, deferred or rejected. Also, the effort required for SCR analysis is estimated.

The first decision point question (SCR accepted?) in the process flow after execution of process P7-2 is reached. If the SCR is rejected or deferred, the answer is no. At the next decision point question (SCR deferred?), if the SCR is rejected, the answer is no, and the flow goes to the Close SCR (P7-6) process. If the SCR is deferred, the answer is yes, and the waiting time countdown to the deferral date starts. At the end of the countdown, the process flow returns to the Review SCR for Evaluation (P7-2) process. A new decision is made whether to proceed with the SCR, defer it to a new date or reject it.

If the SCR is accepted, the answer to the decision point question (SCR accepted?) is yes, and the project stakeholders are notified in the Notify Project Stakeholders (P7-3) process.

The process comes to the next decision point question (small change?). If the estimated time for SCR analysis is under two hours, the answer is yes, and the process flows to the Requirements Frame via entry point 5 for SCR analysis without prior SCR budget approval. Otherwise, the SCR does not qualify as a small change and the answer is no.

The process flow enters the Approve Budget for SCR Analysis (P7-4) process. If the analysis budget is not approved, the answer to the decision point question (analysis budget approved?) is no, and the SCR is closed in the Close SCR (P7-6) process.

If the SCR analysis budget is approved, the answer is yes, and the process flow is directed to the Requirements Frame for SCR analysis via exit point 5. When the SCR analysis is completed, the flow goes back to the Planning Frame via entry point 1B for planning and estimating the SCR and then back to the Scope Change Control (P7) process via entry point 20 to the Approve SCR Implementation (P7-5) process.

If the implementation budget is not approved, the answer to the decision point question (implementation budget approved?) is no, and the SCR is closed in the Close SCR (P7-6) process. Otherwise, the process goes to the Planning Frame for the work breakdown structure updating in order to incorporate the SCR implementation plan into the existing project plan. From there, the flow goes to Construction Frame for SCR implementation and tracking. When the SCR is implemented, the process flow goes back to the Close SCR (P7-6) process to close the SCR via entry point 12.

Submit SCR (P7-1)

When a change to the existing project scope is identified, an SCR is submitted to the project manager, using the SCR form. An SCR must be submitted in order to do any of the following:

- Change one of the existing requirements or design/technology
- Add a new requirement
- Delete an existing requirement
- Change earlier estimates or the schedule, even if existing requirements are not affected

Even if the scope change is small and can be implemented in minutes without additional charge to the client, an SCR still must be submitted in order to avoid potential issues described earlier in this section.

Before an SCR is submitted to a project manager, part 1 of the SCR form (see Table 10.1A) must be filled out by the initiating team lead. Part 1 of the SCR form has the following fields:

1. **Project name**—The officially documented project name.
2. **SCR #**—Because the project manager is the one who keeps track of all sequential SCR numbers, he or she will enter the SCR number after receiving the SCR for review.
3. **Requirement ID**—Reference to the requirement identifier as documented in the traceability matrix. The requirement identifier consists of three parts separated by hyphens: project ID, baseline requirement identifier as documented in the business requirements document and the revision identifier. If, for example, the traceability matrix refers to requirement identifier CLI00253-001-02, where 02 is the last revision identifier, then the new requirement ID will be CLI00253-001-03, where the old revision 02 is replaced with the new revision 03. If the

Table 10.1A Scope Change Request Form Part 1

Scope Change Request	
Project Name:	SCR #: ____________ Requirement ID: ____________
Part 1: SCR Initial Request	
Priority: ☐ Must have ☐ Should have ☐ Nice to have Priority Justification:	
Submitted by (name): ☐ Delivery Team ☐ Business Team ☐ Other Stakeholder	
Short Description of SCR:	
Justification for SCR:	
Dependencies on Other Activities or Projects:	
Impact on Other Activities or Projects:	
Last Update Date:	
Detailed SCR Description (include attachments, if necessary):	
Signature:	Original Issue Date:

SCR is related to a design or technology change, this field should be left blank, because no design elements are included in the traceability matrix.

4. **Priority**—Check the SCR priority (must have, should have or nice to have, as described in the Requirements Frame section of this book).
5. **Submitted by**—Name of the person who submitted the SCR and a corresponding check for the group to which he or she belongs (delivery team, business team or other stakeholder).
6. **Short description of SCR**—A one-line description that provides a hint to what the SCR is about.
7. **Justification for SCR**—Justify the need for the SCR.
8. **Dependencies on other activities or projects**—Another activity or project that must be completed before the SCR can be implemented.
9. **Impact on other activities or projects**—Impact of the SCR on other activities or projects as determined by an impact analysis performed by the requestor of the SCR. It is the project manager's responsibility to ensure that the impact analysis is performed and documented by qualified personnel before beginning the SCR implementation.
10. **Last update date**

11. **Detailed SCR description** (include attachments, if necessary)
12. **Signature**
13. **Original issue date**

Review SCR for Evaluation (P7-2)

When Part 1 of the SCR form has been completed by the requestor and received by the project manager, an identification number is assigned to the SCR. The project manager will review the SCR with the delivery team members and determine whether to accept or reject or to defer review of the SCR until a specified date in the future. The delivery team members will estimate the cost of SCR analysis. The project manager will fill out part 2 of the SCR form (see Table 10.1B) and file it in the project control book.

In those organizations where there is an electronic sign-off process, the SCR is an electronic document, which is electronically signed and distributed. If an electronic sign-off process is not used, then a paper version of the SCR must be completed and signed off on manually. In this case, the electronic copy of the SCR without an electronic signature still must be stored in the project control book, and the project manager must also keep the paper version with the signature. It is also possible to scan the signed copy of the document and store it in the project control book, but the original must still be kept until the end of the project.

Part 2 of the SCR form has the following fields:

1. **SCR accepted**—Checked when the SCR is accepted for further evaluation.
2. **SCR rejected**—Checked if the SCR cannot be justified or is not necessary.
3. **SCR deferred**—Checked when the SCR is put on hold until the date indicated in the "until" field.
4. **Estimated analysis cost**—Estimated cost of the SCR analysis and impact analysis.

Table 10.1B Scope Change Request Form Part 2

Scope Change Request	
Project Name:	SCR #: ____________ Requirement ID: ____________
Part 2: SCR Acceptance for Further Review	
☐ SCR Accepted ☐ SCR Rejected ☐ Deferred until ____________ ☐ SCR Is Small ☐ SCR Is for Design Change Estimated Analysis Cost: ____________ Date: ____________ Project Manager: ____________ Project Manager Signature: ____________	
Comments:	

5. **Date**—Date of SCR acceptance for further evaluation.
6. **Project manager name**
7. **Project manager signature**
8. **Comments**—Detailed explanation of why the SCR was rejected or deferred. Other comments are optional.

All fields are filled out by the project manager. If the SCR is rejected or deferred, the reason must be recorded in the comments field. If the SCR requires further clarification, it must be deferred until such clarification is received.

Notify Project Stakeholders (P7-3)

Project stakeholders are identified and listed during the Communication Management Planning (P6) process. Based on this list, notification of a new SCR will be sent to all people on the list, with a copy of the SCR attached. The following are some key reasons why stakeholders should be aware of every SCR:

1. In any project where several business areas are involved, an SCR initiated by one business area may not be acceptable to other business areas.
2. Often, the number of client-initiated SCRs increases in later stages of the project. This happens because clients can see the product outline better toward the end of the project and want to fine-tune the requirements. Each SCR raises the cost of a project and changes the schedule, and client managers must be fully aware of these changes. This also makes clients provide better justification for each SCR and filters out those changes which are not absolutely necessary.

Approve Budget for SCR Analysis (P7-4)

When parts 1 and 2 of the SCR form have been filled out, the project manager sends the SCR to the client for approval of the SCR analysis budget (unless the analysis effort estimate is less than two hours). The client may accept the cost or reject it. The client fills out the fields in part 3 of the SCR form (see Table 10.1C) and returns it to the project manager.

Part 3 of the SCR form has the following fields:

1. **Approved for analysis**—Checked when the cost of the SCR analysis, which includes an impact analysis, is approved by the client.
2. **Estimated analysis cost**—Estimated cost of the SCR and impact analyses, provided by the project manager.
3. **Analysis rejected**—Checked if the client does not want to proceed any further with the SCR.
4. **Date**—Date of SCR acceptance for further evaluation.
5. **Authorized by**—Name of the person who approved the cost of the analysis.
6. **Signature**

Table 10.1C Scope Change Request Form Part 3

Scope Change Request	
Project Name:	SCR #: ______ Requirement ID: ______
Part 3: SCR Analysis Approval	
☐ Approved for Analysis Estimated Analysis Cost: ______ ☐ Analysis Rejected Date: ______ Project Manager: ______ Authorized by: ______ Signature: ______	
Comments:	

7. **Comments**—Filled out if the SCR analysis is rejected. The reason for rejection must be clearly explained. Other comments are optional.

All fields must be filled out by the authorizing client.

The SCR analysis will also evaluate the impact of implementing the scope change on other project components and deliverables, as well as provide estimates for the SCR analysis. When evaluating the impact, the following must be taken into consideration:

- Project risk, schedule, quality, cost and resources
- Long-term, product-oriented, steady-state product and operations-focused effects
- Other project requirements that may be affected by the SCR
- Changes caused by the SCR to other projects and existing products
- Effect of the SCR on earlier commitments to and by teams and subcontractors affected

Estimates and the impact analysis must be documented in the project control book.

Approve SCR Implementation (P7-5)

The project manager enters the estimated SCR implementation cost in the corresponding field of part 4 of the SCR form (see Table 10.1D), adds his or her name and signature and then sends the SCR to the client for implementation budget approval. If the client approves the budget, then the SCR is sent back to the project manager and the process flow goes to process P5a of the Planning Frame to incorporate the scope change implementation plan into the existing project plan.

Table 10.1D Scope Change Request Form Part 4

Scope Change Request	
Project Name:	SCR #: ____________ Requirement ID: ____________
Part 4: SCR Implementation Approval	
Estimated Implementation Cost: ____________ Estimated Schedule Impact: ____________ Project Manager: ____________ Project Manager Signature: ____________	
☐ Approved for Implementation Date: ____________ ☐ Rejected Name: ____________ Signature: ____________	
Comments:	

Part 4 of the SCR form has the following fields:

1. **Estimated implementation cost**—The estimated cost of the SCR implementation, which also should include costs of other project changes as a result of the SCR implementation.
2. **Estimated schedule impact**—The estimated delay in project implementation due to the SCR. This is not the duration of the SCR implementation, but rather the difference between the planned project duration before the SCR and after the SCR is completed.
3. **Project manager**
4. **Project manager signature**
5. **Approved for implementation**—Checked when the cost of the SCR implementation is approved by the client.
6. **Date**—The approval date.
7. **Rejected**—Checked if the client does not want to implement the SCR.
8. **Name**—The name of the approver.
9. **Signature**—The approver's signature.
10. **Comments**—This field must be filled out if the SCR implementation is rejected. The reason for rejection must be clearly explained. Other comments are optional.

The new SCR status must be documented in the project control book.

Table 10.1E Scope Change Request Form Part 5

Scope Change Request	
Project Name:	SCR #: ______ Requirement ID: ______
Part 5: SCR Closeout	
Changes are complete and SCR is closed. Date: ______ Actual Schedule Impact: ______ Actual Cost: ______ Project Manager: ______ Project Manager Signature: ______	
Comments:	

Close SCR (P7-6)

When the SCR implementation is complete, notification comes from the Construction Frame via entry point 12. The project manager fills out part 5 of the SCR form (see Table 10.1E), indicating SCR completion. Part 5 of the SCR form has the following fields:

1. **Date**—The date of the SCR closeout.
2. **Actual schedule impact**—The actual delay in project implementation due to the SCR.
3. **Actual cost**—The actual cost of SCR planning and implementation.
4. **Project manager**
5. **Project manager signature**
6. **Comments**

The change request is closed when the SCR is completed. All updated documentation is filed in the project control book.

Wait for End of SCR Deferral (P7-7)

This process is the timer set for the duration of the SCR deferral. As soon as the time is up, the process flow returns to process P7-2 to review the SCR again.

Scope Change Process Metrics

The project scope change process uses the following measurements:

1. Plan versus actual date and effort to complete each SCR investigation and implementation

2. Cost of each SCR planning and implementation
3. Impact of each SCR on the schedule
4. Cumulative cost of all SCRs and SCRs grouped by requirements and design changes
5. Cumulative impact on schedule of all SCRs and SCRs grouped by requirements and design changes
6. Number of approved changes to each documented requirement
7. Number of approved design and technology changes

Estimating (P12a and P12b)

Purpose

The purpose of the Estimating (P12a and P12b) process is to describe the steps for developing size, effort, cost, schedule and critical resource estimates for a project throughout its life cycle. Processes P12a and P12b have essentially the same content, but are used for different purposes and are placed in different areas of the Planning Frame.

Accuracy of Estimates

Accuracy of estimates depends on:

- Level of detail, i.e., the degree of decomposition of the work breakdown structure (WBS)
- Risk assessment results and the remediation plan
- Quality of requirements
- The point in the project life cycle when the estimating took place
- The estimator's experience
- The estimating method

There are three levels of estimating accuracy:

1. **Ballpark estimates**—Ballpark or initial estimates are made when little information about the project is available and there are no detailed requirements, except the initial project request. In order to do ballpark estimates, the delivery team must be familiar with similar types of projects and the technology used. This type of estimate is also done when significant risks are involved. The accuracy of ballpark estimates ranges from –25% to +75%.
2. **Preliminary estimates**—Preliminary estimates are performed immediately after completion of business requirements. These estimates heavily depend on the team's familiarity with similar projects, type of business and technology, with no high risks present. A high-level WBS should be used to do this type of estimating. Preliminary estimates are used to establish the preliminary project budget and are often used to establish initial project funding. The accuracy of preliminary estimates never exceeds –10% to +25%.
3. **Accurate estimates**—Accurate or definitive estimates are prepared from well-defined detailed data and the WBS, using the techniques described below. This type of estimate is done just before the project plan package is created or updated. The estimate may not cover the entire project, but instead only the well-defined next stage of the project plan. It is not usually possible to do accurate estimates for an entire project, because of the lack of detailed information for the required activities in the distant future. The best accuracy that can ever be achieved ranges from –5% to +10%.

NOTE In some organizations, where delivery managers with no real project management experience are constantly under pressure from senior management, project managers may face demands for accuracy of estimates better than –5% to +10% or even ±0%. Since this is an unrealistic and unachievable level of accuracy, project managers are forced to use tricks to match real costs to estimates. Because project scope changes are inherent in all projects, one of those tricks is overestimating or underestimating scope changes to keep the visibility of the overall project cost within the required accuracy of estimates in accordance with management's demands. Another trick is using reserve activities for each group of tasks, which are adjusted as necessary to match costs to estimates. In fact, most managers are aware of this, but due to demands from senior management or temptations to report excellent results to the CEO, they keep this practice under wraps. We assert that these tricks provide no real benefits whatsoever and in fact threaten the project and may even cost the project manager his or her job.

Types of Estimates

There are four types of project estimates:

1. **Size and complexity** (software projects only)—Estimates of the project complexity, which are a measure of the project sophistication and the number of inputs and outputs combined with the technology used

2. **Effort**—Estimates of time required to complete tasks outlined in the WBS
3. **Cost**—Estimates based on the effort estimates and resource rates
4. **Critical resources**—Estimates of resources that are needed to support design, development and testing

Documenting Estimates

The project control book documentation should specify how to document and store the estimates and assessment forms related to the Estimating (P12a and P12b) process and other documentation.

Estimates are produced for each new project scope change, for each frame and for the overall project after performing risk assessment and developing risk response plans. Task-level estimates for scope changes and the next upcoming frame should be fairly accurate (–5% to +10%). The overall estimates for the entire project become progressively more accurate with each successive frame of the project. Each time an estimate is created, the following information should be captured in the project control book:

- Date
- Frame and activity
- Inputs (describe the specific inputs used to create the estimate, i.e., requirements document, detailed specifications, design, initial WBS)
- Assumptions made
- Constants
- Estimating method used
- Risk assessment
- Estimates
- Any other information which helped drive the estimates, especially bases for the estimates

Tasks for developing estimates must be included in the project schedule. The Estimating (P12a) process is decomposed into the following subprocesses:

1. Estimate Size (P12-1)
2. Estimate Effort (P12-2)
3. Estimate Cost (P12-3)
4. Estimate Schedule (P12-4)
5. Estimate Critical Resources (P12-5)
6. Review Estimates (P12-6)

In addition, the Estimate Effort (P12-2) and Review Estimates (P12-6) processes are decomposed further. Process P12-2 is broken down into:

1. Perform/Adjust Estimates (P12-2-1)
2. Validate Estimates (P12-2-2)

3. Add Overhead to Estimates (P12-2-3)
4. Perform Alternate Method Estimates (P12-2-4)

Process P12-6 consists of the following low-level processes:

1. Identify Review Team (P12-6-1)
2. Schedule Review (P12-6-2)
3. Conduct Review (P12-6-3)
4. Record Review Notes (P12-6-4)
5. Update Estimates (P12-6-5)
6. Confirm Estimates (P12-6-6)

Estimating Process Flow

Ballpark (or top-down or analogous) estimates involve using past experience and analogy of the initial project request and past projects in the same organization, making adjustments as necessary. If no similar projects were developed in the past, and there are no reliable subject matter experts on the team, it is not possible to provide ballpark estimates.

For ballpark estimates, the process starts when a request comes via entry point 2 to process P12-1 or P12-2 of the Ballpark Estimating (P12b) process. For preliminary or accurate estimates, the Estimating (P12a) process starts when processes P1 to P7, P11 and P14 are complete. If size estimating is the established standard within the organization, then the process flow enters the Estimate Size (P12-1) process. Otherwise, the flow enters the Estimate Effort (P12-2) process. After efforts are estimated in process P12-3, a control point question (ballpark estimates?) is asked. For ballpark estimates, the answer is yes. After producing ballpark estimates, the process flows back to the Requirements Frame via exit point 4. The Estimating (P12a) process flow diagram is shown in Figure 11.1.

If the answer is no, the flow enters the Estimate Critical Resources (P12-5) process. After completion of estimates, a review and approval take place in the Review Estimates (P12-6) process. Once estimates are reviewed and approved, the process flow exits the Estimating (P12a) process for WBS design and update in process P5.

Estimating is done several times throughout the project life cycle:

- Preliminary estimates of the overall project cost and duration produced after the business requirements document is developed—Accuracy of these estimates is –10% to +25%. These estimates may be used to establish initial funding only after performing the initial risk assessment.
- Planning tasks for the next upcoming project frame—Accuracy of these estimates is –5% to +10%.
- Plans are updated by adding a risk contingency as a result of risk assessment.
- Plans are made for a project scope change—Accuracy of the scope change estimates is –5% to +10%.

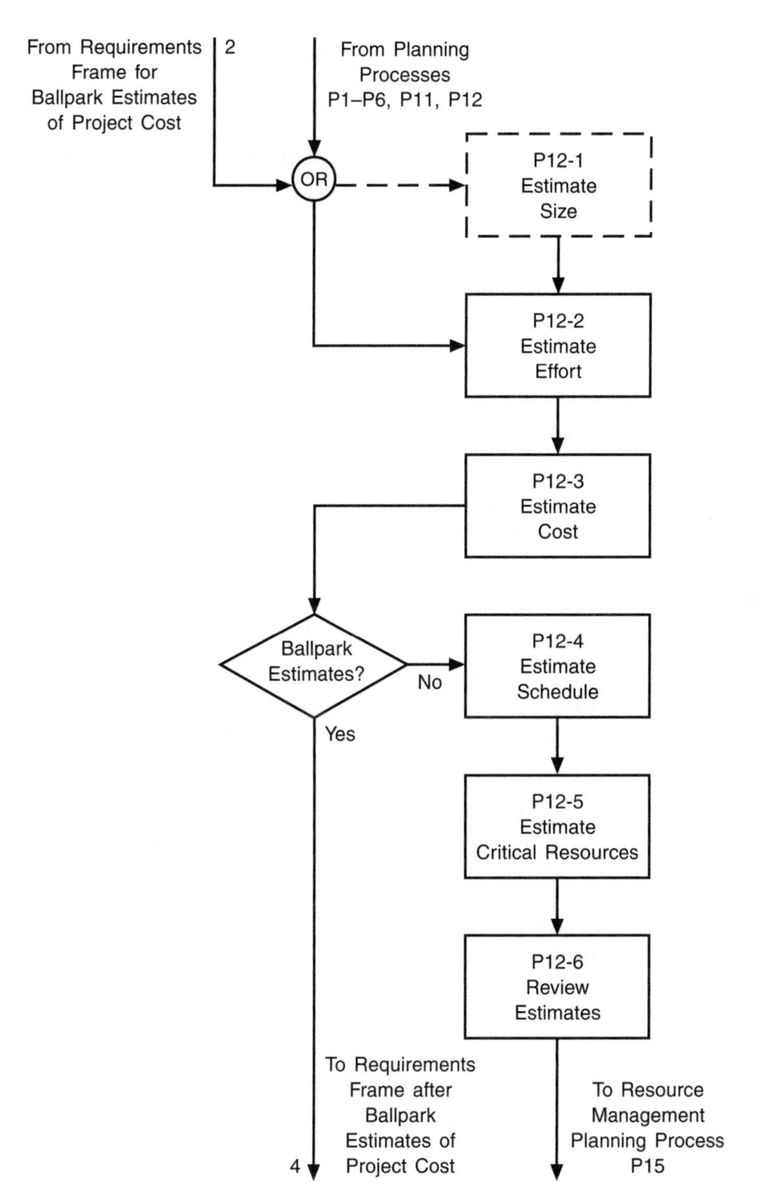

Figure 11.1 Estimating (P12a) Process Flow

- Project overruns the budget or schedule—The accuracy depends on the cause of the poor project performance.

Estimate Size (P12-1)

Size estimates vary with the type of project, because size itself is an attribute of complexity that varies depending on the practice area of the project. The size of documentation or a process may be expressed in standard pages, the size of a presentation may be expressed in number of slides, the size of a bridge may be expressed in terms of tons of steel and the size of a software application may be expressed in number of function points or sometimes in lines of code.

Function points are units of software project complexity, which are associated with the functionality of the software application. They are not directly related to effort required to develop the application, even though such dependency exists indirectly and may be used to rate the developer's efficiency. Thus, if the application size is calculated as 400 function points, then developing that application with MS Visual C++ may take, for example, 6 months using 5 resources with average skill. . However, developing the same functionality using MVS/CICS/Cobol may take 18 months using the same number of resources. Function point analysis, combined with the statistics gathered, allows calculation of the unit cost of software development for every platform being used to develop software. Function point analysis is a method for identifying and classifying components of a system. It measures software by quantifying its functionality based on logical design. Function point analysis may be accurately performed after development of requirements.

Because this type of estimate takes into consideration inputs/outputs, application files, database tables and other parameters, it can be performed for new, not yet developed projects as well as older, completed projects. It is thus possible to create historical data for software developed years ago, provided the actual development efforts were recorded. For example, it is possible to determine that project A, which was developed on a Java platform 3 years ago, has 500 function points and cost $250,000 or $500 per function point. Project B, developed last year also on a Java platform, cost $505 per function point, which is about the same. Therefore, a subcontractor developing a Java project with a cost of $700 per function point is either overcharging, is ineffective in executing the project and/or is using poorly qualified resources. Size estimates for software projects can also be produced using the Constructive Cost Model (COCOMO), where output is measured in lines of code.

Other advantages of using function point analysis include:

- Provides validation of estimates made by the delivery team members
- Controls delivery team productivity
- Improves workload and resource planning

Function point analysis is a complicated method to use. Therefore, only a trained function point analyst can provide function point estimates. A function point count spreadsheet is available free of charge from the International Function Point Users Group and may be used as a tool. Also, there are commercial tools available that automate the process of function point analysis.

However, function points are not applicable to all software activities, because many of those activities, such as software maintenance, redesign for performance improvement and others, do not normally add functionality to existing applications. Therefore, those activities do not add function points, although the effort to perform those activities may be significant.

Use of function points is especially important for offshore software development projects. The most qualified offshore vendors use function points as a standard size-estimating technique without admitting it. You should ask offshore vendors to formally use function points and provide you with results, since a change in an offshore vendor's cost per function point will signal a change in offshore vendor team performance. It also may allow you to evaluate an offshore vendor's performance by comparing it with industry standards or even determine whether using offshore resources provides any real benefit, despite the stated lower hourly rate.

Estimate Effort (P12-2)

The effort, or estimated labor hours, per task may be calculated using one of the following major types of estimating:

- Top down
- Parametric
- Bottom up

Top-Down Effort Estimating

This type of estimating is possible if a sufficient amount of statistical data and experience is available from similar projects developed in the past. Top-down estimating is most suitable for ballpark and preliminary estimates. It is performed by subject matter experts familiar with similar projects. Top-down estimating methods include the following elements:

- **History of past projects and available statistics**—This method is also called the analogous estimating method. Based on past recorded project statistics, the new project is compared with historical data. If similar projects were completed in the past, they are evaluated and the new project is adjusted in accordance with the difference in project scales. Even if the actual work effort was not recorded, it is possible to find it by using the number of delivery team members involved in the project and the length of development.
- **Opinions of subject matter experts**—If team members do not have direct experience in a particular type of project or the technology used, experts outside of the team can be used to provide advice on estimates. They also may be able to help with project implementation.
- **High-level WBS**—A detailed WBS often reaches five or six levels of a tree-like structure. A high-level WBS usually has two and rarely three levels. The lowest level tasks are estimated using the estimating methods described above. A high-level WBS is rarely used for ballpark estimates.

Top-down effort estimating is usually used in projects that do not (yet) have a detailed WBS on a task level.

Parametric Effort Estimating

This estimating method is used by estimating tools that are based on statistical methods, such as Monte Carlo analysis. Those tools calculate the probability of completing tasks within a certain time period based on past experience, by taking advantage of parameters (rates) such as $300 per square foot or 3 engineers per 70-user network design. Sometimes estimates for one or several components may be expanded for the entire project. Parametric estimating is as accurate as the amount of statistical data collected for prior projects. This method is rarely used in software development, but cannot be excluded from consideration. It may potentially be within the range of accurate estimates, even though it is mostly within the range of initial estimating.

Bottom-Up Effort Estimating

This is the most accurate detailed estimating method. Several subject matter experts provide estimates for all tasks in the detailed fully decomposed WBS.

In order to eliminate subjective factors in estimating, the Program Evaluation and Review Technique (PERT) should be used for effort estimating. Each estimator provides three estimates for each task in the WBS: an optimistic (O) estimate, a pessimistic (P) estimate and the most probable (M) estimate. The resulting calculation is:

$$\text{Estimate} = [O + P + (4 * M)]/6$$

Past experience shows that if only one estimate, rather than three, is requested from experts, they most often provide the optimistic estimate. A request to provide three estimates for each task forces them to think about all possible issues in implementing each task. If the results from several estimators differ by more than 5%, their estimates are returned to them along with the results from the other estimators. The experts review the estimates and submit them again. This anonymous, iterative format for honing an estimate is sometimes called the Delphi method.

Estimates from delivery team members usually provide only the net effort required to complete project tasks. Therefore, they are low even if PERT is used, because they overlook daily overhead, such as managing e-mail, meetings and reports. The project manager must add those activities to project estimates. Also, estimates provided by subject matter experts do not take into account resource utilization or results of the project risk assessment. The results of risk assessment may dramatically change estimates or even determine that a project cannot be delivered as planned.

Bottom-up estimating can be performed if a detailed WBS is available from the project planning process. In order to achieve accurate estimates (–5% to +10%), no task in the WBS should require over 40 hours of effort, with best results achieved for tasks of 24 to 36 hours. It is not usually possible to provide that level of detail for the entire project, unless a project is in the late stages of implementation. Only the next upcoming project frame may have accurate estimates after the current frame planning is complete. The later a project is in the project life cycle, the better the overall accuracy of estimates.

In practical terms, the project manager distributes estimating worksheets to the subject matter experts, who provide the lowest level tasks estimates. As mentioned above,

Table 11.1 Task Estimating Worksheet

Project Name: ______________________________

Estimator: ______________________________

Date: ____________________

Check	Task	Optimistic (hours)	Pessimistic (hours)	Most Probable (hours)	PERT Result (hours)
	Activity 1				
	Subactivity Level 1				
	Subactivity Level 2				
✓	Task 1	17	24	22	21.5
	Task 2				
✓	Subtask 1	29	34	31	31.2
✓	Subtask 2	3	7	5	5.0
✓	Task 3	9	15	11	11.3
✓	Task 4	7	10	9	8.8
	Subactivity Level 3				
✓	Task 1	34	39	35	35.5
✓	Task 2	21	30	26	25.8
✓	Task 3	16	22	19	19.0
✓	Task 4	20	27	22	22.5
✓	Task 5	30	38	36	35.3
	Activity 2				
	Subactivity Level 1				
	Subactivity Level 2				
✓	Task 1	14	19	16	16.2
✓	Task 2	15	21	18	18.0
✓	Task 3	14	16		5.0
	Subactivity Level 3				
✓	Task 1	24	35	29	29.2
	Task 2	22	35	31	30.2
✓	Task 3	20	24	23	22.7
✓	Task 4	17	21	18	18.3

no task should exceed 40 hours. Any task that appears to take longer than 40 hours must be further broken down.

Table 11.1 is an example of a task estimating worksheet. Under activity 1, task 1 is the lowest level task and should be estimated. On the other hand, task 2 is further broken down into subtask 1 and subtask 2. Therefore, task 2 is not the lowest level task and should not be estimated; subtask 1 and subtask 2 should be estimated.

Estimate Effort (P12-2) Process Flow

This activity consists of four processes:

1. Perform/Adjust Estimates (P12-2-1)
2. Validate Estimates (P12-2-2)
3. Add Overhead to Estimates (P12-2-3)
4. Perform Alternate Method Estimates (P12-2-4)

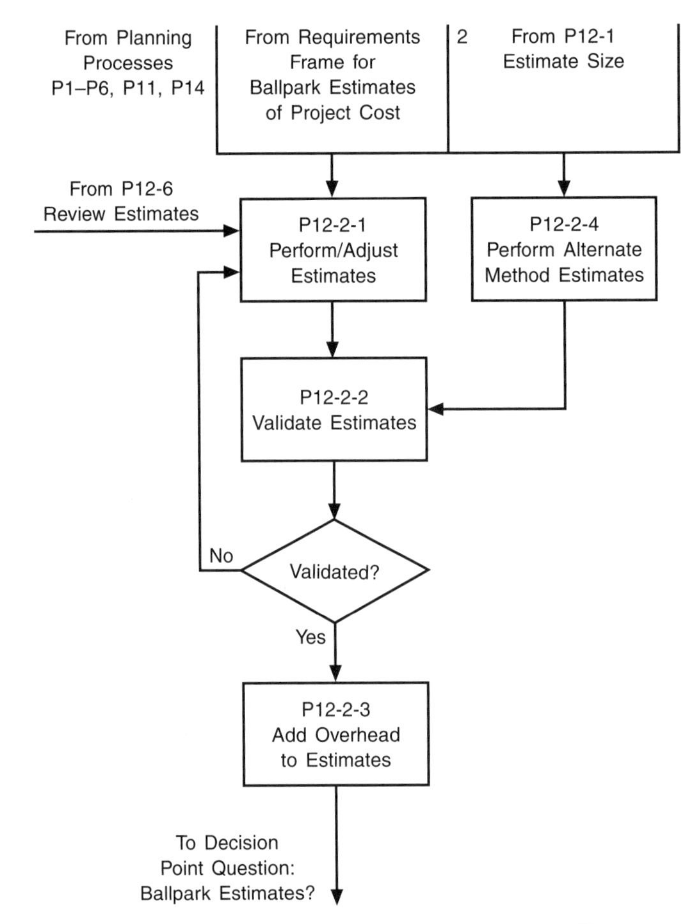

Figure 11.2 Estimate Effort (P12-2) Process Flow

Effort estimates provide labor estimates for the project. The cost and schedule estimates are derived from effort estimates. The Estimate Effort (P12-2) process flow is shown in Figure 11.2.

Perform/Adjust Estimates (P12-2-1)

Project efforts reflect the labor required to complete the project tasks identified. Effort estimates can be done using the methods described above. As the project progresses and more details surface, re-estimating the effort is required using the bottom-up approach. Variances outside the acceptance criteria range, determined during the Construction Frame, will require corrective action. Except for initial estimates, all estimates and adjustments must use the existing project schedule or relevant parts of it. The schedule should be built in accordance with the project planning processes and the list of all available detailed tasks.

The estimates recorded in the task estimating worksheet will be stored in the project control book. Updates are necessary through the project life cycle.

Validate Estimates (P12-2-2)

Effort estimates can be validated in one or more of the following ways:

- Have multiple experts provide estimates, review the estimates anonymously and come to a consensus (the Delphi method)
- Compare results of the main and the alternative estimating methods
- Compare estimates to historical data for similar projects or tasks completed earlier
- Compare the cost of one function point with industry and/or local standards (software projects only)

If effort estimates are not found to be valid, the process flow is returned to the Perform/Adjust Estimates (P12-2-1) process for new estimating or adjustment.

Add Overhead to Estimates (P12-2-3)

Net estimated project effort represents effort to perform tasks initially listed in the task estimating worksheet and later transferred to the project plan. There are, however, additional efforts, which are called unproductive efforts. They may or may not be recorded in the WBS, depending on an organization's methods, tools and processes. Whatever is recorded in the schedule should not be counted again as unproductive time or overhead. For example, if project management tasks, such as developing the WBS, the communication plan and so on, are included in the schedule, the project management effort cannot be added as *additional* effort overhead. Note that it is a good practice to include project management effort in the WBS as a work stream.

The following are components of effort overhead:

- Unproductive time
- Project management and other management activities
- Low resource utilization
- Risk assessment and risk response activities

There is a distinction between effort overhead and cost overhead, which are counted separately.

Unproductive Time Overhead

Delivery team members spend around 20% of their time on phone conversations (some of which may be personal), ad hoc meetings, conversations, breaks, etc.

Project Management and Other Management Activities Overhead

Project management effort tends to comprise between 10 and 20% of the total project effort. Often other managers, like resource managers, line managers and functional managers, charge several hours a week each to the project. This may constitute another 2 to 5% of the total project effort.

Low Resource Utilization Overhead

Resource utilization is the percentage of time that a resource is actually working on the scheduled tasks, as compared to the total time claimed by the resource. It is not possible to achieve 100% resource utilization. This means that if 5 full-time resources are available to work on a project 8 hours a day for 4 weeks, they will charge the project for 800 hours. However, they may accomplish their tasks with a total scheduled effort of much less than 800 hours.

Due to planned task interdependencies, some of the tasks performed by resource B cannot start until other tasks, performed by resource A, are completed, which causes patches of idle time in resource B utilization. To reduce idle time and increase resource utilization, resource leveling must be attempted. Resource leveling is a technique available in almost every automated project scheduling tool to examine an uneven resource load and automatically balance it, *when possible*, also resolving resource overallocation. It does not do well with balancing, but does a good job on overallocation. It is not possible to completely eliminate idle time.

Let's say a resource is assigned for 100% availability to a project for 2 weeks, from Monday through Friday. During this time, the resource completes 3 tasks with a planned total effort of 75 hours. If the resource is not assigned any other work for the remaining 5 hours, he or she will claim 80 hours of labor regardless of the total effort. Therefore, the cost of labor should be calculated for the entire period between the first and the last day of availability, unless the resource does additional work unrelated to the project and charges that time to nonproject activities.

The efficiency of resource utilization depends on the project manager's planning skills. A skilled project manager may reach 90% resource utilization at best. In other words, 10% or more of resource time is often not productive due to inefficiencies in resource utilization. Note that this is in addition to the inefficiencies (such as e-mail and telephone time) described above as unproductive time overhead.

Risk Assessment and Risk Response Activities Overhead

Every project has risks, which add extra effort and extra costs. While costs are derived from effort, there are components of risk that are expressed in terms of cost, such as expected monetary value of risk, and are not direct results of project effort. Those components of risk that require extra effort are risk assessment and planning efforts to reduce or eliminate the probability of a risk occurring and its severity if it does occur. (A risk assessment tool is available for download from J. Ross Publishing from the Web Added Value™ Download Resource Center at www.jrosspub.com or from the authors at www.pm-workflow.com.) Efforts to develop a risk response vary depending on the total number of risks identified and the overall risk rating of the project. The ballpark numbers for risk response planning are 5 hours for medium risk and 16 hours for high risk. The actual implementation of the planned risk response activities may require more time and extra cost. Usually, no risk response is planned for risks with low ratings. If exceptional or high risks are identified, the project should not continue until a new approach is developed and a new risk assessment does not identify those levels of risk.

Table 11.2 Example of Calculating Overhead Effort Estimates

#	Effort	%	Hours
1	Net estimated effort		355.5
2	Unproductive time	20%	71.2
3	Project manager effort	15%	52.9
4	Management effort	5%	17.3
5	Resource utilization overhead	15%	52.9
6	Risk planning and risk response analysis	10%	35.6
	Total effort		585.4

For example, if the delivery team is experienced with only some aspects of the project and not with others, then an assessment must be done to determine the risk of developing each unfamiliar area. If efforts are planned to reduce risk probability and/or severity or to implement the risk response activities for the accepted risks, this may present a significant increase in project cost estimates. The lack of experience may become very expensive to fix if the team has to switch to a different method in the middle of project implementation and redesign some of the project elements. The same must be said about the project manager's lack of experience, which allowed this situation to occur in the middle of the project. Project managers themselves clearly can be a major source of threat to a project!

Example of Calculating Overhead Effort Estimates

An example of calculating overhead effort estimates is shown in Table 11.2.

NOTE

The major issue with the accuracy of project estimates is failure to include at least some of these overhead effort estimates. It is not the responsibility of the project team members, even the most experienced ones, to include the above overhead; rather, it is the project manager's duty to do so.

Perform Alternate Method Estimates (P12-2-4)

An alternate estimating method should be used to develop a second set of estimates. Results of the alternate estimates are used to validate effort estimates in the Validate Estimates (P12-2-2) process. For example, if the main estimating method is bottom-up estimates, then top-down or function point estimates should be done, even if their accuracy is lower than those done using the bottom-up method.

Estimate Cost (P12-3)

Cost estimates are derivatives of effort estimates, where efforts in hours are multiplied by the hourly rate of resources plus additional factors, such as setting up development and test environments, the cost of training, computer equipment, licenses, travel expenses, etc.

Usually, resource rates that project managers receive from management include the cost of resources plus many other indirect costs. Entering rates into project scheduling tools will automatically determine the unadjusted cost of the project. In order to determine the total cost, the following cost adjustments must be made:

- Risk contingency
- Expected monetary value of accepted risks
- Project infrastructure and tools
- Test environment
- Travel
- Training expenses
- Cost of outsourced project activities

Risk management strategies are described in Chapter 6 of this book. The risk contingency fund is allocated to deal with implementation of risk plans to eliminate or contain risks in order to reduce their impact and also to handle unexpected risks.

The goal of risk management is to allow focus on the most important project risks and, after analysis and development of response plans, eliminate or significantly reduce all high and exceptional risks, as well as reduce most of the medium risks to the level of low risks. However, all low and some medium risks are accepted, because the losses due to their occurrence are lower than the cost of preventing them. If they become issues (triggered risks) during the course of the project, they will cause losses equivalent to the expected monetary value of accepted risks. Expected monetary value (= probability * severity) is a statistical assessment of project losses due to risks, not a prediction of final cost. As an example, four risks are identified for a small project in Table 11.3. An example of a risk cost estimate is provided in Table 11.4.

Based on Table 11.3, there should be a risk contingency fund of $7,400 due to the risks identified. Risk #3 is the highest risk, but risking $4,000 in this project is acceptable. However, if the project is much bigger (the cost is around $7 million) and the maximum loss due

Table 11.3 Risk Contingency Calculation Example

#	Risk	Probability of Occurrence (%)	Severity or Maximum Loss ($)	Monetary Value ($)
1	The project schedule is tight but achievable (80% confidence)	20%	$7,000	$1,400
2	There are some doubts about the contractor's timely delivery (bypass is available)	30%	$5,000	$1,500
3	Project manager does not have significant experience and may not be able to deliver on time	40%	$10,000	$4,000
4	Database failure may occur	10%	$5,000	$500
	Total expected monetary value			$7,400

Table 11.4 Risk Cost Estimate Example

#	Item	Cost
1	Total effort (500 hours * $85 per hour)	$42,500
2	Risk contingency	$5,000
3	Expected monetary value of accepted risks	$7,400
4	Project infrastructure and tools	$5,000
5	Travel	$1,000
6	Training expenses	$3,000
7	Cost of outsourced project activities	$20,000
	Total cost	$83,900

to the inexperienced project manager is $1.5 million, then the expected monetary value is $600,000. In this case, consideration should be given to assigning a more experienced project manager. If no local resources are available, an experienced consultant project manager may be hired.

Estimate Schedule (P12-4)

Schedule estimates are based on the WBS after the effort estimates for all tasks, task dependencies and resource assignments have been added to it. All those factors are components of the project schedule, which may be built using scheduling tools, such as MS Project, Clarity, Primavera and others. Methods for building the project schedule are described in Chapter 7. It is assumed that effort overhead has already been taken into consideration when building the project schedule.

Estimate Critical Resources (P12-5)

Critical resources needed to support design and implementation must be identified in this process. Their cost must be added to the project, unless they can be used in other projects, in which case their costs are called capital expenses. For example, the following costs are considered capital expenses:

- Cost of general purpose computers, workstations, desktop computers, laptops, printers, etc.
- Cost of wide- and local-area networks, Internet, etc.
- Cost of heavy equipment in the construction business
- Cost of general purpose machinery to produce mechanical assemblies for the project
- Cost of furniture and buildings

The cost of critical resources must be included in cost estimates, unless it is explicitly stated in the statement of work that those costs are not included in the cost case and must be added separately. Even if responsibility for acquisition of critical resources lies outside of the delivery team, the project manager must identify the critical resources to those who have that responsibility, as the project will be dependent upon them completing their tasks.

The critical computer resource estimates will be documented on the estimating form and stored in the project control book. This information will be communicated to all affected groups.

Notes on Outsourcing Estimating

There is a strong trend today, especially in the software and electronics industries, to outsource parts of projects to offshore vendors. Other than vendor selection, the second most important component of offshore outsourcing is providing the vendor with a carefully developed requirements document. The price quoted by the vendor will be as reliable as the requirements submitted to the vendor. Most failed projects from reputable vendors are due to negligence by the delivery organization in providing vendors with good quality requirements. Should a dispute arise between the parties, a court of law may not find the vendor legally bound to honor the quoted price, even in the case of a fixed price contract, if requirements do not conform to requirements standards established in the organization.

When requirements are forwarded to a vendor, the vendor will be bound by the statement of work, which contains, among other things, the project price quoted by the vendor as a fixed or time and materials price.

There are several variations of a fixed price quote, which may include an incentive clause, a penalty clause and so on, in accordance with the standards and the overall agreement between the delivery organization and the vendor.

A time and materials quote will provide the ballpark or preliminary estimates, even though the actual bills from the vendor will be on the basis of time and materials reimbursement plus a fixed price or fixed percentage, which must be documented in the statement of work.

Time and materials may have some advantages over a fixed price quote from a vendor. In both cases, the vendor will balance the largest possible margin against an attractive price. As more details about the project become available, the actual costs of the time and materials contract may be less than the quote in a fixed price contract.

Review Estimates (P12-6)

The Review Estimates (P12-6) activity is decomposed into the following processes, as shown in Figure 11.3:

1. Identify Review Team (P12-6-1)
2. Schedule Review (P12-6-2)
3. Conduct Review (P12-6-3)
4. Record Review Notes (P12-6-4)
5. Update Estimates (P12-6-5)
6. Confirm Estimates (P12-6-6)

The purpose of this process is to review estimates and ensure that they are complete and ready to be included in the plan and budget for approval.

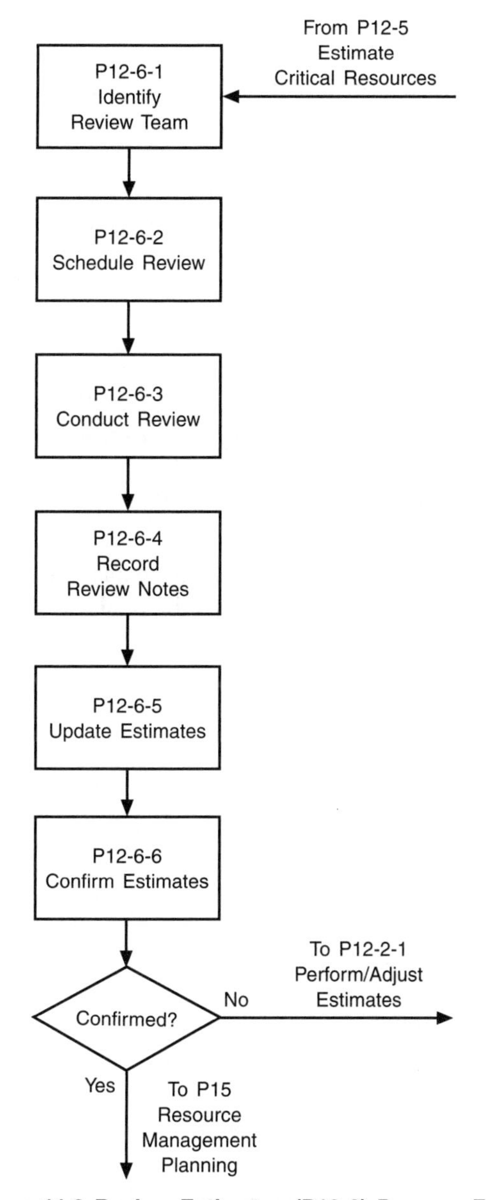

Figure 11.3 Review Estimates (P12-6) Process Flow

Identify Review Team (P12-6-1)

The review team must include the following team members:

- Project manager
- Selected members of the delivery team
- The subject matter experts who provided estimates
- Outsourcing vendor representatives, if outsourcing vendors are involved
- Other personnel as needed

Schedule Review (P12-6-2)

In order to perform a schedule review, at least three days before the review the project manager should:

1. Send review invitations to the participants
2. Send the agenda and review materials to the participants

Conduct Review (P12-6-3)

During the Schedule Review (P12-6-2) process, the project manager should provide participants with the estimates, as well as their validation. The subject matter experts will answer the participants' questions. When reviewing a software project estimate from an offshore vendor, the cost per function point should be compared with industry standards and with size estimates from the same and from other offshore vendors. If the cost per function point differs from industry standards by more than 5%, then an explanation should be requested from the offshore vendor. If the explanation is not acceptable to the participants, the estimate should be returned to the offshore vendor for rework and/or other vendors should be considered.

Record Review Notes (P12-6-4)

The review minutes, including all suggestions and critiques, must be recorded. After the review session is over, the minutes will be sent to the participants for their confirmation that the minutes have been recorded correctly. If necessary, any appropriate corrections will be made.

Update Estimates (P12-6-5)

Estimates can raise contentious issues, and often it is not easy to get 100% consent from everybody. Different participants will have different opinions in accordance with their backgrounds and interests. While some will fight to increase estimates, others will attempt to reduce them. After considering everyone's opinion, the final decision will be made by the project manager, because it is his or her head at stake. The subject matter experts will be asked by the project manager to make corrections to the estimates if necessary. Upon completion, the estimates should be sent back to the project manager.

Confirm Estimates (P12-6-6)

If the project manager is satisfied with the updates to the estimates, the corrections will be sent back to the participants for their information only. It will be up to the project manager to make a final decision based on historical data and confidence in the estimating process and estimators.

Metrics

The following metrics, collected during the Construction Frame, will be gathered during this process:

- Estimates versus actuals for size, effort, cost, schedule and critical resources
- Team performance metrics (based on the effort to deliver one function point if the function point method is used)

The project manager should update the project control book with the above metrics, which will be added to the pool of historical data used for validation of estimates in future projects and for further improvement of the Estimating (P12a and P12b) process.

Completion Criteria

The Estimating (P12a and P12b) process is assumed to be completed when:

- It is identified and documented in the project control book
- All size, effort, cost, critical resource and schedule estimates are documented in the project control book
- Results of estimate reviews and meeting minutes are documented in the project control book
- Approval of estimates by the project manager is documented in the project control book
- Metrics are documented in the project control book

Approve Closing Frame Plan and Budget (P13)

This step is executed here, after planning of the Closing/Testing Frame is complete. In terms of sequence, this happens between the end of the Construction Frame activities and the beginning of the Closing Frame. The final budget determined here has a definitive accuracy of –5% to +10%. The process is initiated by looping back to the beginning of the Planning Frame when the Construction Frame plan is approved.

In order for the plan to be approved, the estimates and the updated project plan, including the user acceptance test schedule, are sent to the client and the senior business manager. By then, the project is largely completed and the only major activity left before closing the project is the user acceptance test. Therefore, it is extremely unlikely that the Closing Frame budget will not be approved. In fact, the process flow does not even consider that option.

12

Outsourcing Management (P14) and Resource Management (P15)

This chapter deals with obtaining and managing resources, both external (P14) and internal (P15).

Outsourcing/Offshore Management Planning (P14)

Purpose

The purpose of the Outsourcing/Offshore Management Planning (P14) process in the context of this book is the selection of qualified vendors to produce quality project components that seamlessly integrate with the other components of the project, as well as managing relationships with those vendors.

Before making a commitment to use outsourcing on a project, there must be reasonable proof that it will be beneficial; there also must be awareness that outsourcing provides significant challenges in producing quality deliverables.

Outsourcing Challenges

Outsourcing, also known as subcontracting, is used in numerous business areas. The government, military, heavy machinery manufacturers and others use onshore resources, as opposed to offshore ones. The heaviest use of offshore outsourcing is in the information technology, electronics and manufacturing industries.

In the past ten years, there has been a strong trend to take advantage of the lower costs of offshore development and manufacturing. This trend takes jobs out of the higher cost country, which is profoundly negative; however, in order to stay competitive, companies cannot afford to ignore this option. In addition, public companies have to follow their shareholders' decisions, which are motivated mostly by profit. It appears that the trend is here to stay for a while, but in the foreseeable future it will slow down or gradually reverse, because the cost of offshore development is steadily going up. As soon as the cost in low-cost countries becomes comparable to the cost of onshore development, the trend will stop. For example, ten years ago, Canada was considered a low-cost country, because the average rate of Canadian developers was about 40% lower than the rate of U.S. developers, in part due to a 40% difference in the value of the Canadian versus the U.S. dollar. Software development by several U.S. corporations was channeled there through their Canadian branches. Today, the cost of software development in Canada is not any lower than in the United States, and Canada is no longer a low-cost country.

Having software development done outside the country presents many challenges. Among them are:

- **Language barrier**—In many countries, English is spoken only by a small minority. This limits communication between the onshore project manager and the offshore team.
- **Time zone difference**—In some of the most popular low-cost outsourcing countries, there is an 8- to 14-hour difference between local time and the U.S. Eastern time zone. This means that during the period between 8:00 a.m. and 5:00 p.m. Eastern time, it may be 8:00 p.m. to 5:00 a.m. in the offshore country, which makes communication for both teams problematic. When e-mail is used for communication, the reply to a query may not be received until the following day. In addition, it is not always easy to get reliable status reports from an offshore project manager until deliverables are due.
- **Different customs and cultures**—In some cultures, people consider asking questions impolite; therefore, business or technical requirements submitted to them must be highly detailed and accurate. If they are not, then the deliverable will reflect the offshore team's understanding and not yours. This sometimes leads to the misconception that offshore teams are not as qualified as U.S. teams, which is generally incorrect. Experience shows that the problems are often caused by poor, or at least vague, specifications and poor communication, rather than a specific problem with the offshore team's qualifications.
- **Privacy and confidentiality issues**—U.S. companies are not allowed to provide offshore companies access to actual production data when it contains private information about customers or other confidential information. This prevents offshore companies from participating in the integration of the deliverables they produce. The data used by the offshore team must be stripped of all references to private and confidential information. Whenever the use of large data banks with millions of records is required to test the system, producing and maintaining a simulated database may be costly and convoluted.

The Outsourcing Process

The outsourcing process consists of the following elements:

1. Issue request for proposal
2. Conduct bidders conference
3. Receive proposals
4. Investigate the candidates' ability to produce the deliverable as an outsourcing subcontractor
5. Select the winning proposal
6. Establish communication and reporting with the outsourcing company
7. Provide the outsourcing company with the detailed statement of work and specifications
8. Receive the project plan from the outsourcing company
9. Track implementation
10. Track integration with other parts of the project

When outsourcing help is required, the company issues a request for proposal. For onshore outsourcing, the announcement may be published in a newspaper or sent to a list of potential bidders. For offshore outsourcing, the request for proposal is sent to a list of known and reputable offshore companies doing outsourcing business. Questions may be answered in a bidders conference, which may be either a face-to-face meeting or a teleconference with bidders that expressed interest. When proposals are received by the due date, they are reviewed, and several are selected for further consideration. The bidders whose proposals were selected are investigated for financial strength and availability of the appropriate resources. All references should be checked. There must be sufficient evidence that the outsourcing company has delivered successfully on similar types of projects in the past with satisfactory quality. Based on the investigation and the suitability of the proposal, a winning proposal is selected. There must be a documented understanding with the outsourcing company that its project management processes will be adjusted to make them compatible with standard processes in the contracting organization.

It is rarely a project manager's responsibility to select outsourcing companies. Most often, particularly in large organizations, a special procurement unit is responsible for selection of offshore vendors. Those vendors work with multiple project managers in the organization, and there is no other choice but to use their available resources. A documented process for selection of outsourcing vendors should be established in every organization. Its implementation must be strictly controlled by senior managers, because business ethics of overseas companies may differ.

When the outsourcing company is selected, the communication process must guide the relationship, as described in Chapter 9. Sometimes, instead of dealing with the outsourcing vendor's project manager, a local resident technical representative is assigned. This improves communication to a certain extent, but the representative cannot fully speak for the project manager and does not have authority to make important decisions.

The next step in the process is development of the detailed statement of work, as outlined in Chapter 13, which is signed off on by representatives of both sides. A detailed specification must be included in it or provided separately to the outsourcing organization. Based on the statement of work and other documentation submitted to the outsourcing company, the outsourcing project manager must develop and present his or her project plan for approval to the onshore project manager. No work should be started by the outsourcing organization without a signed-off statement of work and an approved project plan. Some organizations use a work authorization form to formally authorize the start of project work, with reference to the detailed work specifications, the authorized budget and method of payment, as well as a list of people authorized to maintain contact between both organizations.

All tracking activities are usually performed by the outsourcing project manager and reported weekly, as specified by the Communication Management Planning (P6) process. If issues come up, they will be resolved in accordance with the Issue Management Planning (P4) process. The quality of deliverables must be controlled by the Quality Management Planning (P2) process, described in Chapter 14.

When deliverables are complete, they must pass quality control reviews by the onshore company's quality management department, just as if they were developed onshore. The acceptance criteria, which are described in the statement of work, must be used for approval of deliverables.

The Outsourcing Plan

The outsourcing plan consists of the outsourcing process description, which is recorded in the project control book, and the scheduled outsourcing activities integrated into the overall project plan.

Resource Management Planning (P15)

Purpose

The purpose of human resource management is managing resources on the project. The effective use of resources is a key contributor to project success, since resources, or more likely the *lack* of them, directly influence project costs, duration and quality. Resource Management Planning (P15) consists of the following processes, as shown in Figure 12.1:

1. Plan Resources (P15-1)
2. Acquire Resources (P15-2)
3. Assign Resources to Tasks (P15-3)
4. Build the Team (P15-4)

Resource planning should start as early as possible to account for the time needed to actually obtain resources. The resource plan is created early in the project, but is updated every time the project plan is updated. The more we learn about the project and

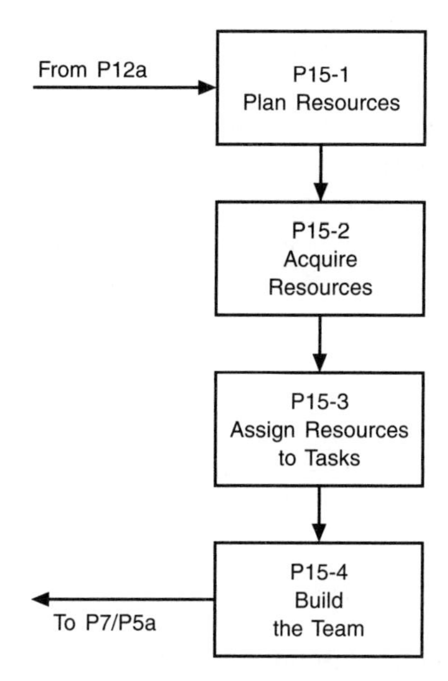

Figure 12.1 Resource Management Planning (P15) Process Flow

the more detailed the project plan becomes, the better idea we have about needs for human resources, thus allowing more accurate resource estimates to be produced.

As the project proceeds through the life cycle, both skills and resources go through significant variations. Figure 12.2 shows that at the start of the Requirements Frame, during initiation, the project has few resources, which increase in number during project planning, then increase again in the Construction Frame and then reduce rapidly as the project moves toward the end. At the end of the project, all remaining resources are released.

After receiving estimates for all tasks in the project work breakdown structure, resources for implementation of those tasks should be assigned, as described in Chapter 7. At that time, the actual resource names may not be available, so each task is assigned a resource by skill, rather than a person's name, such as designer 1, designer 2, program-

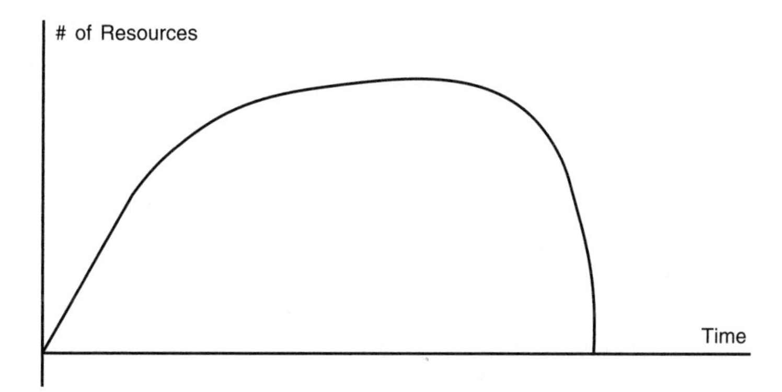

Figure 12.2 Project Life Cycle Resource Distribution

mer 1, engineer 1, etc. This will allow the project manager to have a good understanding of the number of resources required, the level of their proficiency in the required skills and the period of time when each resource is required. The overall project cost depends on the cost of project resources.

Obviously, resource utilization efficiency will affect resource requirements. The project manager should also take into account the average resource's sick leave and vacation time.

Plan Resources (P15-1)

The resource plan consists of the resource management process description, which is recorded in the project control book, and the acquisition resource activities schedule, which may be embedded in the overall project plan.

Acquire Resources (P15-2)

In a matrix organization, human resources are often owned by a resource manager. The resource manager is in charge of hiring, firing, promotions, rewards, resource placement, etc. When any of those actions are required, the project manager coordinates these activities with the resource manager. However, the daily management of resource activities on the project is handled by the project manager. Any less authority of a project manager over resources is counterproductive and puts project success in question.

Resource acquisition is a difficult task, requiring excellent negotiation skills. Resource availability depends on the established priority of projects in the organization. If all resources are engaged in higher priority projects, there is nothing the project manager can do to obtain resources except work behind the scenes with senior stakeholders to raise his or her project's priority. If this cannot be done, then management will reassign the project manager to a higher priority project. The issue of resource acquisition will not disappear, albeit it will be for a new project.

Before requesting resources for the project, the project manager must identify the resource managers who can provide resources with the required skills. A project manager who is new to the organization should speak to his or her manager, fellow project managers and the client to get this information and then make a 30-minute appointment with each resource manager for a preliminary discussion about availability of skills before formally requesting resources. It is a good idea to prepare a 10-minute presentation (the proverbial "elevator speech") about the project, mentioning names of the most influential stakeholders. This may help. The project manager should be friendly with the resource managers and find additional topics of conversation beyond a resource request.

If still no resources are available, there are three options that must be fully explored with the resource managers, delivery manager and client:

1. **Acquire resources shared with other equal priority projects**—The resource load of the assigned resources will be less than 100% and the work duration will be longer than the effort planned for the task. While the cost of the task will not increase, the schedule will change, which may increase the overall project cost to

Table 12.1 Resource Request Form

Resource Request					
Project Name: ______			Project Manager: ______		
Resource Manager: ______			Date: ______		
#	**Resource Type**	**Skill Description**	**Start Date**	**End Date**	**Resource Name**
1	Programmer 1	2 years C++	02/01	04/30	John Smart
2	Programmer 2	3 years Java	02/01	04/15	Fred Little
3	Programmer 3	2 years Java	02/15	03/20	Ann Goldsmith
Resource Request Approved on (date): ______. Resources are identified.					
Resource Manager: ______					
Comments:					

the client. In order to determine the impact on the project schedule and cost, the project plan must be updated to reflect the new reality.

2. **Acquire project resources outside of the organization**—This means that resource managers may hire a new employee or a temporary consultant. Hiring consultants may be more expensive than using company employees. Also, motivation for a consultant to put in extra effort on the project is often limited to extra compensation.
3. **Postpone project implementation until the resources become available**—This decision will require lowering the project's priority and must be made by senior company management.

Once skills availability is confirmed, the resource request form shown in Table 12.1 must be filled out and submitted to each resource manager. The resource manager must enter resource names for each skill identified. If resources have planned time off during the period of assignment, this must be indicated as a comment and the end date extended accordingly.

For example, if a programmer resource with three years of Java programming language is required, but a resource with only two years is available, then additional training will be required and should be scheduled by the project manager. Training is rarely paid for by clients, so the cost of the training should not appear in the project budget. However, the schedule may be affected, depending on the time required for training.

Assign Resources to Tasks (P15-3)

Upon receiving resource approvals, the project manager will substitute the generic resource name, like programmer 1, with the actual resource name. Also, the resource rate must be entered in the scheduling tool, which allows for project financial management and tracking, as described in Chapter 15.

Build the Team (P15-4)

Team building is often described as getting the right resources not just on hand but *onboard.* Unless you work for a global enterprise with resources all over the world, you will not have much choice when resources are scarce. In that environment, the well-regarded team members are snapped up by other projects with more influential stakeholders. Often, you are happy to have *any* resource that even remotely satisfies the required skills.

Lacking a manager's authority to hire, fire, reward and promote team members, the project manager has to build trust and be a motivator. Trust is always mutual. To build trust, the project manager must trust the team and always keep his or her word. The project manager must not be afraid to support team members and to stand up for them, even when that is not too popular with management.

The project manager must create a collaborative environment, where each team member has a leading role in at least one project activity. For example, if several team members work on related tasks, the leading authority to integrate those tasks is given to one team member. The authority for issue resolution during integration should be given to different team members, so that everybody is given responsibility and has accountability, which comes with responsibility. This way, the team members develop mutual respect.

Social events play an important role in team building. The project manager must plan for social activities which are enjoyed by the vast majority of team members. The project manager should try to make sure that if some team members do not participate in one event, they will join the next time around. One good approach is for the whole team to go out after work once in a while and share some food and conversation. The project manager should try very hard to get funding for such events. If that is not possible, the cost may be shared by team members or funded by the project manager. Building the team involves many other forms of human interaction, including formal and informal regular meetings, e-mails, teleconferences, informal gatherings, etc.

Motivating the Team

Motivation is a driving factor in team performance. There are several theories of motivation. The most practical one is the Expectancy Theory of Motivation, developed by Dr. Victor Vroom, according to which people will be productive and motivated if three conditions are satisfied:

1. People believe that their efforts will likely lead to successful results
2. People believe that successful results will be rewarded by a salary raise, benefits or other expressions of appreciation
3. The rewards are important to them

In the successful team, people regard the project manager's success as their own. Tension in the relationship between the project manager and team members presents a barrier to motivation. The project manager's management style—and attitude—play an important role in project success.

Project Management Style

The project manager has a certain amount of authority in the organization to get things done. This authority comes from the fact that the project manager is authorized by his or her organization to manage the project and the team (usually documented in the project charter). This authority may be strengthened by the project manager's expertise in the technical field of the project, even though this commodity is becoming rare today. If the project manager has influence with senior management and is able to act as a "figurehead" for the team and insist on rewarding team members for their effort, this helps ensure respect for that project manager.

An undesirable substitute for reward is when a project manager's demands are enforced with the help of a resource manager. If team members receive nothing for good performance but are still punished for poor performance, this reduces motivation and creates a feeling of animosity. Team members may find satisfaction in poor project performance and, as a result, weakening the project manager's status within the company. While using punishment is not a good management style, it cannot be totally discounted. It should be used with team members who have a poor attitude, when very few other options are available.

Finally, some project managers who feel a lack of respect from team members choose to avoid asking them to do certain things and instead cite other people's authority to request those things be done. For example, a project manager may say that a senior manager wants something done. This type of management style further weakens the project manager's authority and should be avoided.

Conflict Resolution

Conflicts in the project team are inherent. After all, projects are about change, and people do not generally like change. That is one source of conflict. Then there is the fact that the resources report elsewhere and not to the project manager. That is another source of conflict. On top of that is the pressure of working in the unknown, since all projects are unique. Add to this the fact that the team is also working on a tight schedule and budget, and yes, that is a recipe for quite a bit of conflict. Some conflicts are created as a result of a personal dislike between team members, but most often they are due to differences in approach to various project solutions.

Resolving conflict may help both sides to understand and agree on the best solution and take a straightforward approach in solving the conflict. Here, both sides present their opinions, or use the confrontation of ideas in a problem-solving way, avoiding personal confrontation. For the project manager, we recommend three key ideas which come from the area of integrative bargaining:

1. Gather and work with facts, facts, and more facts
2. Focus on interests, not positions
3. Separate the people from the problem

If necessary, arbitration by subject matter experts may be used to choose the best solution. In some cases, a combination of elements from all ideas, or compromise, may be reached, which results in partial satisfaction of both sides without a clear winner. It is important to distinguish between a compromise and an attempt to appease both sides by ignoring differences without resolving the conflict (smoothing), which may flare up again soon. In some cases, when the tension is very high, it can be useful for one or both sides to temporarily stop addressing the conflict in order to let both sides calm down. When the tension subsides, the conflict may be addressed again in the confrontation of ideas mode.

The most undesirable way to resolve conflict is forcing one side to stop bringing the issue up. The losing side will shut up, but will not forgive, and a new conflict will be just around the corner.

13

Statement of Work (P10)

This chapter includes, along with the Create Statement of Work (P10) process, the High-Level Design/Architecture (P11), the Create/Update Plan Package (P8) and the Approve Construction Frame Plan and Budget (P9) processes, because they are precursors to the statement of work.

High-Level Design/Architecture (P11)

A high-level design provides the overall project design and the overall architecture of the solution without getting into implementation details. This process is executed at the same time as the rest of the Planning Frame processes (P1, P2, P3, P5, P6, P14). The flow for process P11 will vary for different industries.

Create/Update Plan Package (P8)

The project plan package is a collection of various subsidiary plans and documents required for project or partial project budget approval. The plan usually contains the following information:

1. Project schedule with all tasks and their dependencies included, resources assigned and estimates made
2. Quality management plan
3. Risk management plan
4. Resource management plan

5. Communication plan
6. Configuration management plan
7. Other plans as needed

Not all these plans are separate documents. If the data required by a plan is included in the project schedule, then no separate document need be produced. For example, if all quality management audits and reviews are scheduled as required by the quality management process, then no separate quality management plan should be produced. However, it is assumed that all required checklists and templates established in the Quality Management Planning (P2) process are documented and used as described in Chapter 14.

The project plan package is first assembled after project planning takes place. It is usually updated when project scope changes are planned, significant project risks are found, the project overruns the budget/schedule, etc.

By the time the process flow gets to the Create/Update Plan Package (P8) process, all required plans should be available. This process consists of bundling all project plans together in one package.

Approve Construction Frame Plan and Budget (P9)

This step is executed at the end of the Planning Frame, just before the Construction Frame starts. The final budget is determined at this point, with a definitive accuracy of –5% to +10%. The budget and the project plan package are sent to the client and the senior business manager for approval. When the budget is approved or approval is withheld due to clearly specified reasons, the process ends.

Create Statement of Work (P10)

This process describes the steps necessary to create the statement of work (SOW), which is the single most important legal document, as it lays the foundation for the project.

Purpose

The purpose of the SOW is to outline the work required for project completion. The SOW helps to control client expectations. It is written and finalized during the Planning Frame of the project. Contents of the SOW include:

1. Revisions
2. Overview and objectives
3. Assumptions
4. Scope
5. Approach

6. Major deliverables
7. Major milestones
8. Delivery organization responsibility
9. Client responsibility
10. Ownership and license
11. Out of scope
12. Completion criteria
13. Project benefits
14. Charges and payments schedule
15. Approvals page

The description of the contents of the SOW indicates the general format of the SOW, which documents the relationship between the client and the project owner (delivery organization). The project owner is responsible for client satisfaction, managing client expectations and implementation of project requirements.

Revisions

This section lists all revisions to the document, dates of approval and names of approvers. Table 13.1 is a sample document revision form. An additional signature page at the end of the SOW will be added for each new revision. Project scope changes, as described in Chapter 10, do not require changing the SOW, but do include modifications to the traceability matrix.

Overview and Objectives

This section provides a brief overview of the project and the attached business requirements definition document. It should state that the terms and conditions described in the SOW are agreed upon by both parties for the period of the specific beginning and ending dates of the engagement. Roles and responsibilities of both sides and their representatives must be explicitly described. All changes to the SOW must follow the existing change control process and an additional sign-off page must be added to the SOW upon accepting changes. It should be stated that the delivery organization will maintain an agreed upon level of quality of deliverables and all other SOW objectives. The client must agree that the existing project management processes of the delivery organization will be followed.

Table 13.1 Document Revision Form

#	Revision	Date	Approved by	Description

Assumptions

All project assumptions must be listed here. Some general examples of assumptions are:

- The delivery organization will provide qualified staffing to perform the agreed upon work on the agreed upon schedule.
- The delivery organization will comply with the established processes of the project sponsor, as well as with all mandatory local and national laws and regulations and with other relevant requirements unless agreed otherwise.
- Each of the individual efforts described in the SOW will be staffed by the delivery organization. Any additional work will be subject to a separate SOW.

Scope

The delivery organization will manage the following project activities:

- Development of each deliverable, listed separately
- Management of all project activities

Approach

Here the overall approach to implementation and the technology used must be described. If integration with other project modules or projects is required, it must be described here.

Major Deliverables

The major deliverables to the client must be described here:

Client Deliverables
  Milestones or the project plan
  Test plan objectives
  Integration plan
  ....
Architecture
Deliverable 1
Deliverable 2
....

Major Milestones

The major schedule milestones and delivery dates will be specified as follows:

Milestone 1	mm/dd/yyyy
Milestone 2	mm/dd/yyyy
Milestone 3	mm/dd/yyyy

Milestone 4	mm/dd/yyyy
Milestone 5	mm/dd/yyyy
Milestone 6	mm/dd/yyyy

It is important to keep these at a high enough, reasonable level that conveys meaningful information but prevents overdisbursement of detailed information.

Delivery Organization Responsibility

The following are some examples of delivery organization responsibilities to be highlighted in the SOW:

- Assign project manager to manage the project staff, to maintain communication with the client and to be the focal point for all communication with the client.
- Assign other members of the project staff.
- Provide the best effort to deliver the project at the agreed upon schedule and budget.
- Provide the client with status reports as documented in the communication management plan.
- Maintain client satisfaction.
- Ensure that qualified resources are assigned to project tasks in the most effective way.
- Monitor actual monthly charges once the project has started, and work with the client to reconcile and validate variance from the original plan.
- Maintain all quality management activities in accordance with the project quality management plan.
- Assist the client in conducting user acceptance testing. The acceptance criteria must be in conformance with the agreed upon specifications as documented in the project control book.

Client Responsibility

- Assign a focal point, who will be responsible for daily communication with the delivery team and for the overall business direction of the project.
- Provide the delivery team with responses to queries within 48 hours and sign-offs on requested documentation within 72 hours or as mutually agreed otherwise. If sign-off is denied due to issues with quality, defects, requirements or understanding of the requirements, the reply must state the exact reason why. If the response, sign-off or denial is not received during this time frame, the delivery team will continue the work as previously scheduled and all charges will be incurred by the client.
- Ensure that funding for the agreed upon work is available.
- Participate in all project status meetings as described in the communication plan and review status reports weekly, providing feedback when needed.
- Provide payment for the work performed on a monthly basis or as mutually agreed and documented.

- Authorize travel of delivery team personnel when required.
- Provide any agreed to supporting facilities and services required by delivery team personnel.
- Advise the delivery team in advance of all changes and modifications that may impact the implementation of the SOW.
- Strictly follow the change control process of the delivery team.

Ownership and License

The ownership of the intellectual property produced by the delivery team during the project belongs to the delivery team. The client and its customers and agents are granted full user license to this intellectual property.

Out of Scope

Everything that is not explicitly documented will be assumed to be out of scope. The basis of the documentation is the original baseline requirements plus all approved change requests.

Completion Criteria

These criteria take precedence over all other criteria in any documentation. The SOW will be considered to be complete when the *first* of the following occurs:

1. All deliverables described have been produced and approved by the client.
2. The delivery team has completed the work under the SOW and has accomplished the tasks and provided the deliverables, which the client did not explicitly reject by sending appropriate notification as described in the client responsibility section.
3. The client has decided to cancel the contract and not complete the balance of the project work. In this case, the delivery team must be notified 45 days in advance, and the client will be charged for any costs incurred, unless staff members can be successfully reallocated to other projects.

Charges and Payments Schedule

The delivery team will invoice the client monthly for professional fees based upon *actual* time worked on the engagement until the end of the project at the rate specified in the contract. Table 13.2 shows an example of monthly charges.

The calculations in Table 13.2 are based on actual rates with a 20% contingency added for the delivery organization efforts.

Direct project-related travel and living expenses for the project team members will be billed separately at actual cost.

Approvals Page

An example of an approvals page is provided in Table 13.3.

Table 13.2 Example of Monthly Charges

Charges for the Month of January 2014		
Labor Description	**Hours**	**Cost**
Effort	1,200	$84,000
Management overhead	60	$580
Total		$84,580

Table 13.3 Example of Approvals Page

Project (or Statement of Work Revision) Authorization	
Client (provide name): Address: Phone:	
Delivery Organization (provide name): Address: Phone:	
Project Name: Estimated Project Start Date: Estimated Project End Date: Estimated Engagement Costs: $	
The *delivery organization* will provide and the *client* will accept professional services for the *project* under this Statement of Work. The objectives, scope, deliverables, *delivery organization's* responsibilities, *client's* responsibilities, completion criteria, and other applicable terms are described on the preceding pages. The parties acknowledge that they have read this Statement of Work, understand it, and agree to be bound by its terms and conditions. Both parties to this agreement accept this contract and agree to all terms and conditions denoted in this Statement of Work.	
Delivery Organization:	
Name, Position	Date
Client:	
Name, Position	Date

14

Quality Management (P2)

Purpose

The Quality Management Planning (P2) process produces a set of planned and executed project activities to ensure that all project deliverables conform to the documented and agreed upon requirements and meet documented stakeholder expectations.

Quality management controls all project management processes described in this book and ensures the project has the required quality. The main vehicle to achieve that target is the quality audits and reviews which are planned and executed. All noncompliance issues found must be resolved by the project team or escalated to senior management in those cases where local resolution is not possible. If several different departments or organizations are involved in the project, the set of quality management processes must be recognized and implemented on an enterprise-wide level. As mentioned in Section II on the Requirements Frame, quality is conformance to the documented and approved requirements and specifications, and therefore it is not the same as the "perfect" performance which clients and business users would like to have. Quality is not a part of any project variable, unlike cost, time and resources. Quality can never be compromised and it cannot be changed under any circumstance. If the project duration, budget or resources are not realistic to implement all business requirements, then some requirements should be dropped or changed to reflect the new situation as mutually agreed with the client. *This newly defined set of requirements will become a new quality standard for the project.* Still, all deliverables must be produced with the agreed upon quality—no less and no more.

No variations in quality may ever exist in the deliverable, since the documented quality is either *there* or it is *not there.* Therefore, it is not possible to exceed the deliverable quality expectations, because that would mean the product has unplanned, unbudgeted elements of scope and does not conform to the agreed upon specifications. The project may be ahead of or behind the schedule and may be below or above the budget, but it is impossible for it to be of better or lesser quality.

Quality management of a project means building and implementing plans to meet the documented quality expectations of clients and stakeholders. The planning effort contains two major groups of processes:

1. Quality Assurance Audit (processes P2-1-1 through P2-1-7)
2. Quality Control Review (processes P2-2-1 through P2-2-7)

There is a major difference between quality assurance and quality control. Quality assurance focuses on the environment in which deliverables are created and the guiding processes. In fact, the key distinguishing aspect of quality assurance is that it is all about process. Quality control, on the other hand, focuses on the inspection of deliverables and removal of defects to ensure the deliverables are complete and satisfy the stated quality. The terms quality assurance and quality control cannot be used interchangeably.

Quality management activities start at the beginning of the Requirements Frame and end after the project is delivered to client at the end of the Closing Frame.

Quality Assurance

The purpose of quality assurance (QA) is to build quality into deliverables from day one of the project, rather than to inspect finished products. QA is all about project processes. The major vehicle of QA activities is the QA audit. The QA audit will judge the quality of the future deliverables *based on the quality of the processes used for their creation.* QA audits are conducted by a QA analyst, with active participation of the project manager and his or her team. The auditor does not have to know the project well, but must be trained to ensure project compliance with processes such as requirements management, risk management, scope change control, project tracking and all other applicable processes which *build the quality into the project.* Even if a project is on schedule and within the budget, poor results of QA audits will credibly predict a troubled project in the near future, unless quality gaps are quickly eliminated.

When senior managers receive project documentation for sign-off, usually they cannot rely on their firsthand knowledge of project details, but rather sign off based on indirect circumstantial evidence of quality. If, let us say, the senior business manager is required to sign off on the business requirements document, he or she will probably attempt to verify with the lead client that the following is true:

- Business users from all the relevant business areas participated in gathering requirements and agree with the business requirements document.
- The lead client signed off on the document.
- The document is clear and unambiguous and contains only necessary requirements.

- The delivery and business organizations are comfortable with the proposed delivery content, the schedule and the project estimates.
- Risk assessment has been recently conducted and the conclusion is that the overall project risk is low.
- The business requirements document review was successful, with no open issues.

If this is indeed the case, the business manager will sign off on the business requirements document without personally going through the entire document.

QA audits are conducted many times throughout a project in accordance with the schedule developed (and in particular the quality management plan). The bigger the project, the more quality audit activities required.

Ideally, the QA department is an independent unit in charge of all company QA activities. It schedules QA audits of all projects and sends QA analysts to conduct them. However, if a QA department does not exist, the role of the QA analyst may be played by a QA-trained analyst of the delivery organization who is not a team member of the project being reviewed.

Except for small projects under $100,000, most QA activities are led by a QA analyst and performed by all members of the delivery team. In small projects, the QA analyst's function may be delegated to a QA-trained project manager.

All QA audits must be planned. While the QA analyst is expected to propose the QA audit plan, it is the project manager's responsibility to create a QA plan and document it in the project control book. The quality plan identifies quality standards and ways to satisfy them. Development of the quality plan includes:

- Definition of all types of QA audits
- Development of checklists for all types of QA audits
- Established frequency of each QA audit type
- Assigned QA resources
- Scheduled QA audits
- Scheduled QA training

QA audits evaluate project performance on a regular basis to ensure that the project satisfies quality standards. QA analysts start conducting QA audits no later than four weeks after the beginning of a project and then at least quarterly. Project audits also take place when new high risks are encountered and when there are issues with cost or schedule (more than 10% variance between actuals and the plan). The QA audit reports are forwarded to the QA manager, if that position exists, and also to senior delivery managers. In addition to QA audits, QA analysts verify that the project runs as planned and all required project documentation and other deliverables are produced in accordance with the schedule. Evidence of QA audits must be recorded in the project control book. Each project deliverable must also pass a separate quality control review, as described later, in order to remove defects and confirm its correctness and fitness for the purpose. All elements of noncompliance must be reported to executive management.

During the monthly executive project status review, as described in Chapter 9, the project manager reports to executives on QA project activities. This information must be reviewed and initialed by the QA analyst prior to the executive project status review.

The main vehicle for QA audits is checklists. Checklists contain questions that must be answered yes or no. If the answer is yes, proof of the answer, such as meeting minutes, memos, documents, e-mails and protocols, must be documented in the project control book. If an answer is no, then there is a gap, which must be closed in order to get a passing grade. In order to pass an audit, all questions in the checklists must be answered yes. For example, if the answer to the checklist question "Does the client understand that even small scope changes will likely affect project cost and schedule?" is yes, proof must be provided in the form of the meeting minutes with the client where this was clearly explained and agreed upon.

Each audit will be rated pass or fail. In the case of failure, the comments section of each audit checklist must explain the reason for failure and include a specific list of actions to rectify the situation. Before forwarding results of audits to the project manager and the senior delivery manager, each audit must be signed by the QA analyst. There must also be proof that the project manager received the results. In the case of a deficiency, a deadline must be established to fix the noncompliance. If the issue is not fixed by the deadline, it must be escalated by the QA manager to the senior delivery manager.

QA audit documentation consists of a front page and a set of checklists (described below). All applicable checklists must be used. The QA Audit (P2-1) process consists of the following components:

1. Inform Project Manager of QA Audit and Send Checklists (P2-1-1)
2. Answer Checklist Questions (P2-1-2)
3. Schedule QA Audit (P2-1-3)
4. Conduct QA Audit (P2-1-4)
5. Eliminate Gaps (P2-1-5)
6. Rate QA Audit (P2-1-6)
7. Escalate Noncompliance (P2-1-7)

Quality Assurance Audit (P2-1) Process Flow

The QA Audit (P2-1) process flow is shown in Figure 14.1. A QA audit is triggered by the QA plan, schedule or when significant issues are found in the project. Approximately two weeks in advance of a scheduled audit, the QA analyst sends a reminder to the project manager, as indicated in the process Inform Project Manager of QA Audit and Send Checklists (P2-1-1). Along with the reminder, all relevant checklist forms must be sent to the project manager, who in the process Answer Checklist Questions (P2-1-2) answers all questions. Checklists are only considered complete when answers to all questions are yes and documented proof is available. Any question answered no means that a gap between current practices and expectations exists. Any waiver to this requirement, as an exception, must be by mutual consent between the QA manager and the

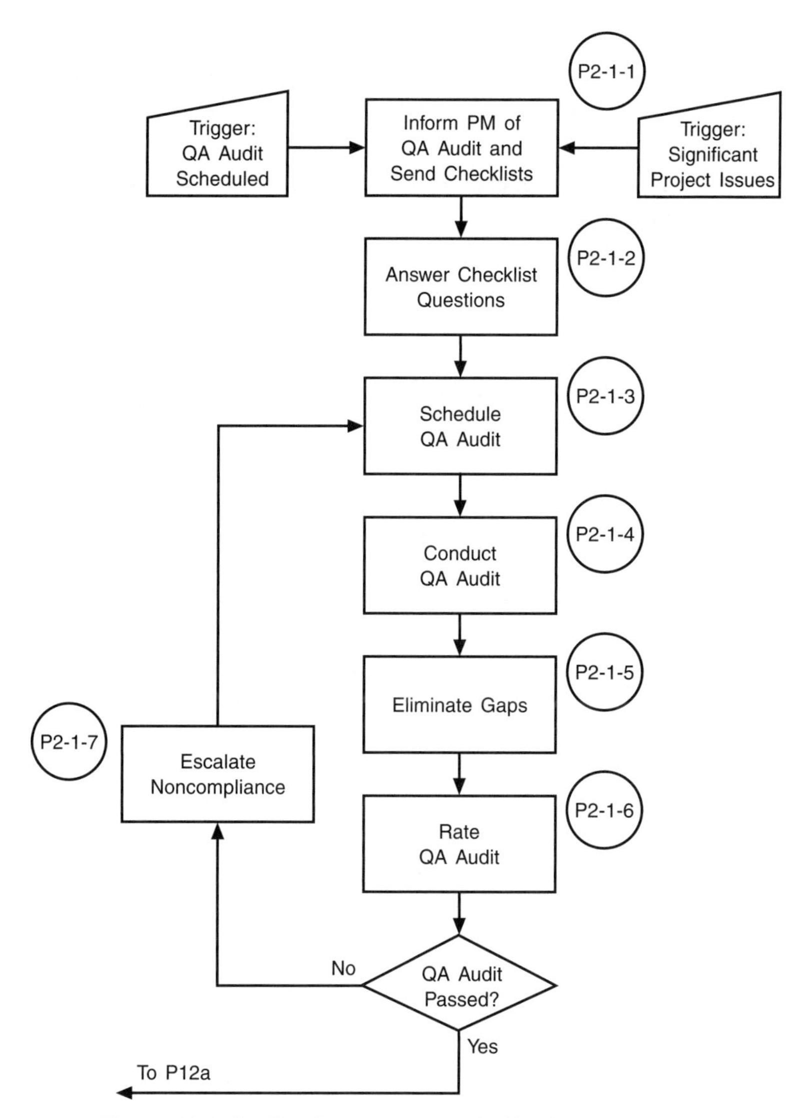

Figure 14.1 Quality Assurance Audit (P2-1) Process Flow

senior delivery manager and documented in the project control book. All gaps must be eliminated by the agreed upon deadline.

Approximately one week before the audit, in the process Schedule QA Audit (P2-1-3), the QA analyst sends an invitation to the project manager for a QA audit at a specified place and time. Often the audit venue is the project manager's office, because most project documentation can be accessed from there.

At the scheduled date and time, the QA analyst and the project manager go through all the checklists, one question at a time, in the process Conduct QA Audit (P2-1-4). Other delivery team members, while not usually present during the entire audit, may be called in and questioned. The project manager must provide a detailed explanation of all answers. If the QA analyst is convinced that the yes answers are correct, it is not always necessary to request documented proof for all answers, even though two or three answers in each checklist must still be verified. If doubt exists, the documented proof must be presented and delivery team members may be questioned. If issues are found during the audit, the QA analyst must provide details in the comments section of the corresponding checklist. At the end of the audit, the project manager is provided with a copy of all checklists, while originals are kept by the QA analyst.

After the audit, the project manager usually has three to four days, or as agreed otherwise, to fix all issues of noncompliance in the process Eliminate Gaps (P2-1-5). The project manager must prepare a document that describes measures taken to eliminate all gaps and forward it to the QA analyst.

After review of the modified checklists and the corresponding documentation, the QA analyst issues the audit rating in the process Rate QA Audit (P2-1-6). If the QA analyst is satisfied with the results, the rating is pass. The answer to the control point question (QA audit passed?) is yes, and the audit is considered complete. The process flow enters the Estimating (P12a) process. The rating must be documented in the project control book.

If all gaps were not eliminated as required, the audit rating is fail, and the answer to the control point question (QA audit passed?) is no. The report must be forwarded by the QA analyst to the QA manager in the process Escalate Noncompliance (P2-1-7). The QA manager takes the issue to the senior delivery management team. If no QA manager position exists, then the report is forwarded directly to the senior delivery manager. Management will review the reasons for noncompliance and take corrective action. When all issues are resolved, the QA analyst is notified, and a new QA audit is scheduled.

Quality Assurance Audit Front Page

The 11 types of QA audits are as follows:

1. Project management process
2. Statement of work
3. Client satisfaction/involvement
4. Financial
5. Change control
6. Staffing
7. Communication
8. Project planning
9. Project tracking
10. Outsourcing/offshore management
11. Configuration management

Table 14.1 Front Page of Quality Assurance Audit

Quality Assurance Audit	
QA Analyst:	
Project Name:	
Project Manager:	
Lead Client:	
Budgeted Project Cost:	
Estimated Project Cost:	
Last Risk Assessment Rating:	Date of Risk Assessment:
Audit Types (check all applicable): ☐ Project Management Process ☐ Statement of Work ☐ Client Satisfaction/Involvement ☐ Financial ☐ Change Control ☐ Staffing ☐ Communication ☐ Project Planning ☐ Project Tracking ☐ Outsourcing/Offshore Mgmt. ☐ Configuration Mgmt.	
Audit Date: ______________ Total Checklists in the Audit: ________ Audit Rating: ☐ Pass ☐ Fail QA Analyst Signature: ______________	
Repeat Audit Date: ______________ Total Checklists in the Audit: ________ Repeat Audit Rating: ☐ Pass ☐ Fail QA Analyst Signature: ______________	
Next Scheduled Audit Date:	
Comments:	

The front page of the QA audit form is shown in Table 14.1. All the information requested in the top section must be provided. The box for each type of audit completed must be checked. The QA analyst completes the section for each audit and gives a pass or fail rating. Comments are noted at the bottom.

Project Management Process Audit

The project management process audit is an overall audit which ensures that the project is managed in accordance with the existing mandatory project management processes and QA. The overall project management checklist (see Table 14.2) is used to conduct the audit. The checklist also confirms that all project costs are taken into consideration.

Table 14.2 Overall Project Management Audit Checklist Form

Project Management Process Audit Checklist		
Project Name: ______ Date: ______		
Project Manager: ______ Client: ______		
		Yes/No
1	Is there compliance with all mandatory project management processes on the organization level?	
2	Are delivery team members familiar with the project management processes?	
3	Does the project proceed in accordance with the detailed project plan?	
4	Is the project stable, with no major scope changes expected?	
5	Does the client understand that even a small scope change will likely affect project cost and schedule?	
6	Is the project within the budget and on schedule (variance is under 10% of schedule and cost) and no foreseeable delays are expected?	
7	Is the cost of QA activities included in the project cost?	
8	Is the cost of risk management activities included in the project cost?	
9	Does the latest project status report reflect up-to-date information?	
10	Is there a project control book, in which ALL project-related information is stored?	
11	Have the project manager and delivery team members had QA training?	
12	Does the project control book contain QA training materials and the minutes of the QA training for the project manager and the delivery team?	
13	Has a QA plan for the project, which includes audit dates, been developed?	
14	Are there records in the project control book regarding all QA activities and all metrics?	
15	Is the traceability matrix updated with every scope change?	
Comments:		

Statement of Work Audit

The statement of work audit confirms that all business requirements can be implemented using the selected methods and technology at the price quoted according to the time frame and milestones indicated. The statement of work audit checklist (see Table 14.3) is used for the audit. This review takes place after the statement of work is completed, but before signing off on it. The review is attended by the QA analyst, project manager, relevant members of the delivery team and subject matter experts. The statement of work audit takes place once, before delivery of the statement of work to the client.

Table 14.3 Statement of Work Audit Checklist Form

Statement of Work Audit Checklist		
Project Name: ______ Date: ______		
Project Manager: ______ Client: ______		
		Yes/No
1	Do the statement of work guidelines exist and is following them mandatory?	
2	Will the proposed solution satisfy all business requirements?	
3	Was a peer review of the proposed implementation conducted?	
4	Was a cost estimate produced and verified by other estimating methods?	
5	Was the price quoted to the client approved by a qualified financial analyst?	
6	Is the milestone implementation schedule reasonable and imposes no undue hardship on the delivery team?	
7	Are the statement of work terms and conditions clear and complete?	
8	Are the correct assumptions made?	
9	Are all legalities taken into consideration?	
Comments:		

Client Satisfaction/Involvement Audit

Managing client expectations is one of the project manager's responsibilities. In doing so, it is important that client expectations be based only on existing project documentation, such as the signed-off business requirements document and the project plan. The client satisfaction rating is based on the client satisfaction survey tool (see Figure 14.2) completed by the client (the tool is available for download from J. Ross Publishing from the Web Added Value™ Download Resource Center at www.jrosspub.com and from the authors at www.pm-workflow.com). A rating under 85% should be considered unsatisfactory. The client involvement audit checklist (see Table 14.4) is used to conduct the audit.

Financial Audit

The financial audit evaluates project performance using earned value analysis, which can be performed using the earned value analysis tool shown in Figure 14.3 (available for download from J. Ross Publishing from the Web Added Value™ Download Resource Center at www.jrosspub.com and from the authors at www.pm-workflow.com). In the tool, budget at completion, actual cost, earned value and planned value must be entered, and cost variance, schedule variance, cost performance index, schedule performance index, estimate at completion and variance at completion are automatically calculated.

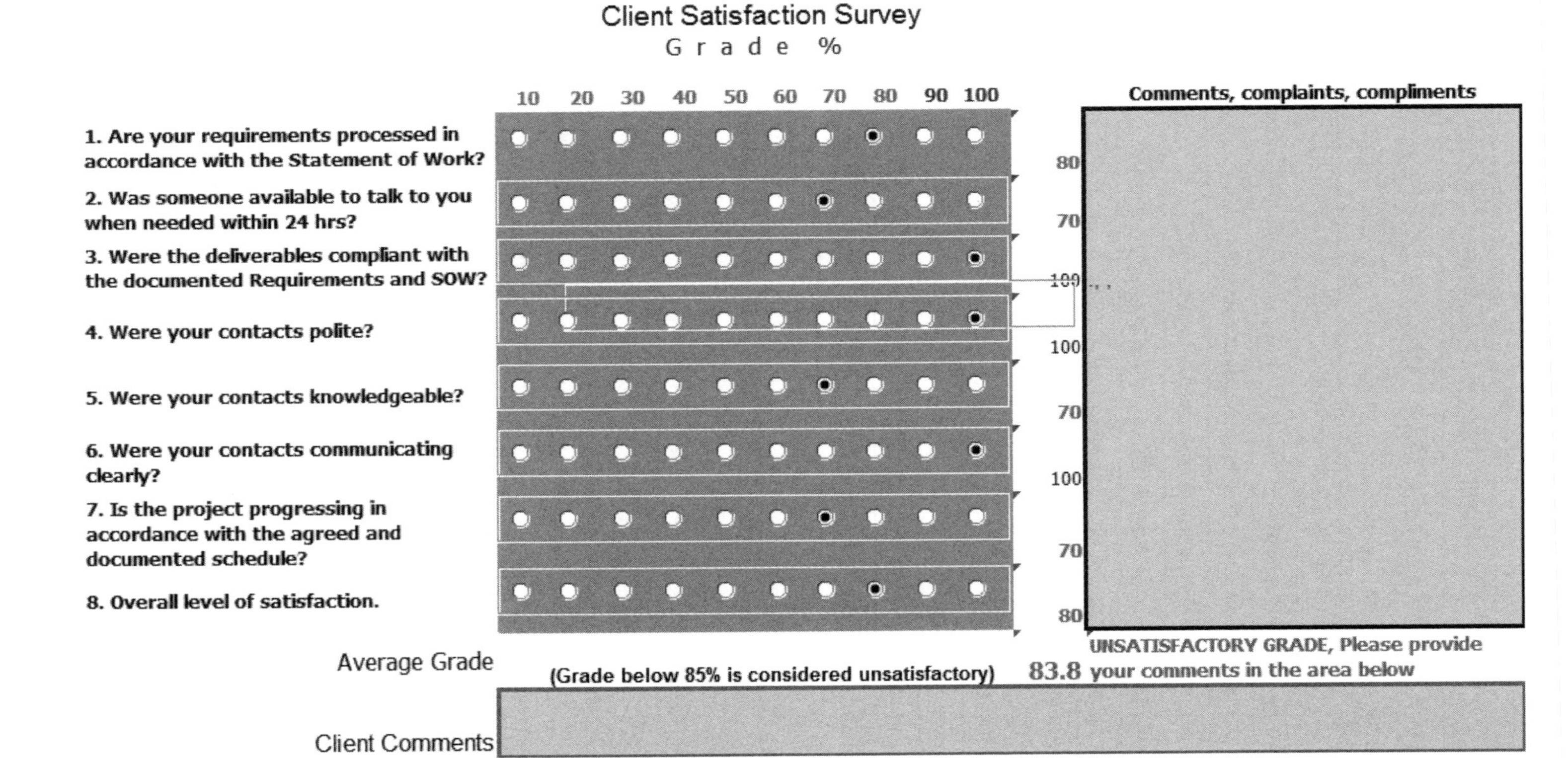

Figure 14.2 Screenshot of Client Satisfaction Survey Tool

Table 14.4 Client Involvement Audit Checklist Form

Client Involvement Audit Checklist		
Project Name: ____________ Date: ____________		
Project Manager: ____________ Client: ____________		
		Yes/No
1	Is the client actively involved with the project?	
2	Are there weekly project status meetings with the client?	
3	Do the delivery team and the client have good business relations?	
4	Did you expect customer satisfaction with the delivery team performance to be at least 85%?	
Comments:		

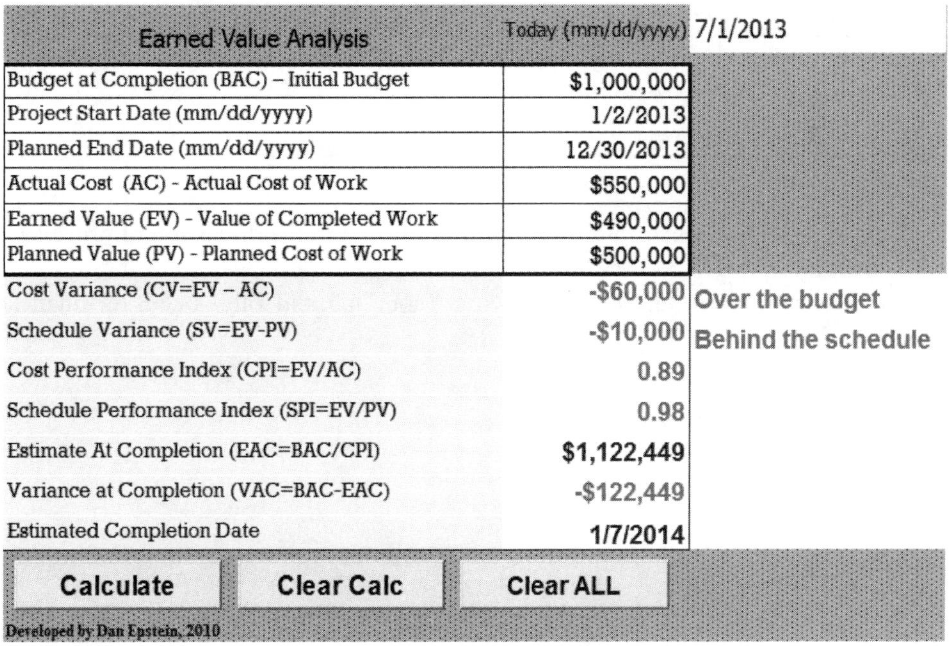

Earned Value Analysis	Today (mm/dd/yyyy)	7/1/2013
Budget at Completion (BAC) – Initial Budget	$1,000,000	
Project Start Date (mm/dd/yyyy)	1/2/2013	
Planned End Date (mm/dd/yyyy)	12/30/2013	
Actual Cost (AC) - Actual Cost of Work	$550,000	
Earned Value (EV) - Value of Completed Work	$490,000	
Planned Value (PV) - Planned Cost of Work	$500,000	
Cost Variance (CV=EV – AC)	-$60,000	Over the budget
Schedule Variance (SV=EV-PV)	-$10,000	Behind the schedule
Cost Performance Index (CPI=EV/AC)	0.89	
Schedule Performance Index (SPI=EV/PV)	0.98	
Estimate At Completion (EAC=BAC/CPI)	$1,122,449	
Variance at Completion (VAC=BAC-EAC)	-$122,449	
Estimated Completion Date	1/7/2014	

Calculate | Clear Calc | Clear ALL

Developed by Dan Epstein, 2010

Figure 14.3 Screenshot of Earned Value Analysis Tool

Table 14.5 Financial Audit Results Form

Financial Audit Results		
Project Name: ______	Date: ______	
Project Manager: ______	Client: ______	
Budget at Completion: ______	Actual Cost: ______	
Earned Value: ______	Planned Value: ______	
1	Cost Variance	
2	Schedule Variance	
3	Cost Performance Index	
4	Schedule Performance Index	
5	Estimate at Completion	
6	Variance at Completion	
7	Overall Project Cost Status (1 = over budget, 2 = under budget, 3 = on budget)	
8	Overall Project Schedule Status (1 = behind schedule, 2 = on schedule, 3 = ahead of schedule)	
Comments:		

Further details about earned value analysis are provided in the discussion of process C1 in Chapter 15.

The financial audit results form is shown in Table 14.5, and Table 14.6 is the financial audit checklist.

Change Control Audit

The change control audit checklist is shown in Table 14.7.

Staffing Audit

The staffing audit checklist is displayed in Table 14.8.

Communication Audit

The communication audit checklist is shown in Table 14.9.

Table 14.6 Financial Audit Checklist Form

Financial Audit Checklist		
Project Name: ____________ Date: ____________		
Project Manager: ____________ Client: ____________		
		Yes/No
1	Is the project running on schedule within the approved budget (up to 10% schedule and cost variance)?	
2	Are the actual efforts spent captured at least weekly?	
3	Are the captured actual efforts compared to planned ones at least weekly?	
4	Are other costs, such as travel, equipment, etc., captured at least weekly?	
5	Is earned value analysis performed at least weekly?	
6	Have approved rates been established, periodically reviewed and used for cost calculations?	
7	Have the costs of approved scope changes, new issues and risks been captured and the budget adjusted accordingly?	
8	Is the approved budget adjustment included in the new baseline?	
9	In the case of a troubled project, are steps taken to ensure the budget will be sufficient to correct the situation and complete the project as agreed?	
10	Is the overall cost of the project adjusted in accordance with the total number of decision-making clients who are allowed to forward their decisions directly to the project manager? (10% cost increase for every additional decision-making client over one)	
Comments:		

Project Planning Audit

The project planning audit checklist is displayed in Table 14.10.

Project Tracking Audit

The project tracking audit checklist is shown in Table 14.11.

Outsourcing/Offshore Management Audit

The outsourcing/offshore management audit checklist is shown in Table 14.12.

Table 14.7 Change Control Audit Checklist Form

Change Control Audit Checklist		
Project Name: ____________ Date: ____________		
Project Manager: ____________ Client: ____________		
		Yes/No
1	Does a change control process exist and is following it mandatory?	
2	Are all the change requests submitted documented?	
3	Are all the approved change requests documented?	
4	Are all the rejected change requests documented?	
5	Are all the implemented changes approved?	
6	Are the change management statistics documented?	
7	Are all the implemented scope changes reflected in the traceability matrix?	
Comments:		

Table 14.8 Staffing Audit Checklist Form

Staffing Audit Checklist		
Project Name: ____________ Date: ____________		
Project Manager: ____________ Client: ____________		
		Yes/No
1	Is a project staffing plan available?	
2	Have all documented staffing requests been approved by resource managers?	
3	Are there any unresolved staffing issues?	
4	Are the skills and resources of the delivery team sufficient for the project?	
5	Is the performance of the delivery team members within industry standards?	
6	Are the skills and resources of outsourcing teams sufficient? (if applicable)	
7	Is the performance of outsourcing team members within industry standards? (if applicable)	
Comments:		

Table 14.9 Communication Audit Checklist Form

Communication Audit Checklist		
Project Name: ______ Date: ______		
Project Manager: ______ Client: ______		
		Yes/No
1	Does a project communication process exist and is following it mandatory?	
2	Is there a documented communication plan?	
3	Is there a documented list of all project stakeholders and their contact information?	
4	Is a single focal point available for contact with the client organization?	
5	Does a weekly status report meeting with the client take place?	
6	Does a monthly status report meeting with the senior business manager take place?	
7	Does a monthly status report meeting with delivery management take place?	
8	Are there any client expectations beyond those documented in the statement of work?	
9	Is a financial report included in the monthly meeting with delivery management?	
10	Is a QA report included in the monthly meeting with delivery management?	
11	Is the client available within 24 hours to discuss urgent issues?	
12	Does the client provide answers and sign-offs on time?	
13	Does the client strictly adhere to the scope change process?	
14	Are the hours reported weekly by the delivery team approved by the client on time?	
Comments:		

Configuration Management Audit

The configuration management audit checklist is shown in Table 14.13.

Quality Control

The purpose of quality control (QC) is verification of compliance of the deliverables to specifications and removal of defects that remain in the finished deliverables. However, if quality is not built into the deliverables from day one of the project using QA, then it is past the time to speak about project quality—all that can be done now is to remove

Table 14.10 Project Planning Audit Checklist Form

Project Planning Audit Checklist		
Project Name: ____________ Date: ____________		
Project Manager: ____________ Client: ____________		
		Yes/No
1	Does a project planning process exist and is following it mandatory?	
2	Is there a documented risk management plan and regularly scheduled risk assessments?	
3	Are the QA plan and regular QA audits included in the project plan?	
4	Are QC reviews included in the project plan and conducted for each deliverable?	
5	Does an estimating process exist and is it used for estimating?	
6	Are estimates created in the estimating process verified?	
7	Do the delivery team members agree with the estimates?	
8	Are the project estimates documented?	
9	Are all work breakdown structure tasks under 40 hours?	
10	Are the risk contingency plans documented and included in the project plan and overall estimates?	
11	Are all approved change request implementation plans made and entered on time in the overall project plan?	
12	Have all task dependencies been identified and entered in the work breakdown structure?	
13	Have available and qualified resources been assigned to all project tasks?	
14	Are the milestones significant to the client included in the work breakdown structure?	
Comments:		

obvious defects. In that case, QC success will be limited, because it is impossible to remove all defects, and some will even be discovered later, in steady-state use and long after project completion.

Deliverables in the project context are not only those elements of the project that are delivered to the client, but also other "intermediate" project elements, which are required in order to produce client deliverables. Therefore, a deliverable is the product of a project task, which has a completion deadline in the schedule.

There are four standard QC reviews:

Table 14.11 Project Tracking Audit Checklist Form

Project Tracking Audit Checklist		
Project Name: ____________ Date: ____________		
Project Manager: ____________ Client: ____________		
		Yes/No
1	Does a project tracking process exist and is following it mandatory?	
2	Has a project scheduling and tracking tool, such as MS Project, Niku, Primavera, etc., been used on the project?	
3	Do all delivery team members provide a weekly status report on the work assigned to them?	
4	Is the project plan updated weekly to reflect the actual hours spent on completed tasks and tasks being processed?	
5	Is a weekly delivery team status meeting scheduled and conducted?	
6	Is presence at the delivery team status meeting compulsory for all team members?	
7	Is the cost of the weekly delivery team status meeting included in the overall project estimates?	
8	In the case of an issue, is corrective action taken and reported by the scheduled date?	
9	Are actuals collected, documented and used to determine the project status at least weekly?	
10	Are all completed project tasks listed in the work breakdown structure?	
11	Is an escalation path established and used for delivery team issues that cannot be resolved by the due date?	
Comments:		

1. Business requirements review
2. Statement of work review
3. Other deliverable documents reviews
4. Peer reviews

The QC Review (P2-2) process consists of the following components:

1. Identify Review Team (P2-2-1)
2. Schedule Review and Invite Participants (P2-2-2)
3. Send Materials to Participants (P2-2-3)

Table 14.12 Outsourcing/Offshore Management Audit Checklist Form

Outsourcing/Offshore Management Audit Checklist		
Project Name: ____________ Date: ____________		
Project Manager: ____________ Client: ____________		
		Yes/No
1	Has selection of outsourcing organizations been competitive and are they thoroughly researched before granting contracts?	
2	Does a contract or statement of work for each outsourcing organization exist, which defines the exact scope of requirements, all deliverables, schedule and cost?	
3	Is a separate business requirements document attached to the statement of work?	
4	Is there a statement of work prepared by the delivery team and approved by QC review?	
5	In the case of a scope change required later in the project, is it reviewed by the delivery team and approved by QC review before being submitted to the outsourcing organization?	
6	Are terms of payment and legal responsibilities thoroughly listed in the statement of work?	
7	Are the outsourcing organization project plan and deliverables approved by the delivery project manager?	
8	Are effective project management and QA processes established and used in outsourcing organizations on all projects?	
9	Has a project scheduling and tracking tool, such as MS Project, Niku, Primavera, etc., been used by an outsourcing organization on the project?	
10	Does the outsourcing organization's project manager provide weekly status reports on the work?	
11	Is the outsourcing project plan updated weekly to reflect the actual hours spent on completed tasks and tasks being processed?	
12	Is the weekly outsourcing status meeting scheduled and conducted?	
13	In the case of an outsourcing organization issue, is corrective action taken, reported and resolved by the assigned deadline?	
14	Are actuals collected, documented and used to determine outsourced project status at least weekly?	
15	Is an escalation path established and used for outsourcing issues that cannot be resolved by the due date?	
Comments:		

Table 14.13 Configuration Management Audit Checklist Form

Configuration Management Audit Checklist		
Project Name: ____________ Date: ____________		
Project Manager: ____________ Client: ____________		
		Yes/No
1	Does a configuration management process exist and is following it mandatory?	
2	Is responsibility assigned to someone for storage of all intangible deliverables in the special computer library and tangible deliverables in special storage?	
3	Does the configuration management computer library and/or storage exist?	
4	Is checking in and checking out of every deliverable documented in the project control book and performed by authorized personnel in accordance with the configuration management process?	
5	Are all configuration-management-related metrics recorded and available for configuration management audit?	
6	Is the cost of configuration management activities included in the overall project cost?	
Comments:		

4. Conduct Review/Take Notes (P2-2-4)
5. Update Materials (P2-2-5)
6. Rate QC Review (P2-2-6)
7. Modify Deliverable (P2-2-7)

The business requirements review was covered in Chapter 3. In addition to the reviews listed above, there are two more QC reviews of deliverables, which are described in Chapter 16:

1. User acceptance test plan review
2. Project completion review

Table 14.14 shows three types of QC reviews, with a brief description of each, timing of the review, participants and frequency. Each QC review is performed once for each deliverable, provided the review is successful, but must be repeated in the case of failure. Depending on the type of project, additional project documentation may be produced which requires review as well. For example, in projects where a proposal was developed, a proposal review is required.

Table 14.14 Three Types of Quality Control Reviews

Review Type	Verify	Time of Review	QC Team	Frequency
Statement of work and initial risk assessment	1. All business requirements can be met 2. Pricing and terms of payment are appropriate 3. Schedule is achievable 4. Legal issues have been considered 5. Overall project risk is low	Before statement of work is delivered to client	Project manager, delivery management, QA analyst, subject matter expert(s)	Once
Other deliverable documents and risk assessments	Each document has different verification criteria specific to the document reviewed	After document is complete, but before delivery	Project manager, QA analyst, subject matter expert(s)	Once for each deliverable document
Peer review	1. The project elements and each deliverable reviewed conform to specifications 2. Defects are removed	After project elements are complete, but before delivery	Project manager, QA analyst, subject matter expert(s), delivery team members	Once for each deliverable

Peer review is usually a technical review of specific project deliverables. There may be one peer review of the project architecture document, one for the high-level design, one for each module of the detailed design, and so on.

The process of producing project deliverables, such as business requirements, the statement of work, the high-level design and others, is part of the project management process, examined by the QA audit, but the actual review of those deliverable documents with the purpose of confirmation and defect removal is a part of the QC process.

Most QC review processes are very similar from the standpoint of preparation for the review, defect removal and review rating. The document walk-through, however, is very specific for each project, with a different approach for each document and its unique audience. The approach will be established by the project manager with the help of the subject matter experts.

QC reviews are led by the project manager or the subject matter expert(s), with the participation of the QA analyst. In addition, someone who is sufficiently knowledgeable about the subjects discussed at the review must be assigned to document all discussions and suggestions.

Quality Control Review (P2-2) Process Flow

A business requirements review, which is one of the QC reviews, is conducted when the Approve Requirements (R8) process is executed in the Requirements Frame. Since requirements management is a "mini-project" in itself, this quality review was described earlier and therefore is not repeated in this chapter.

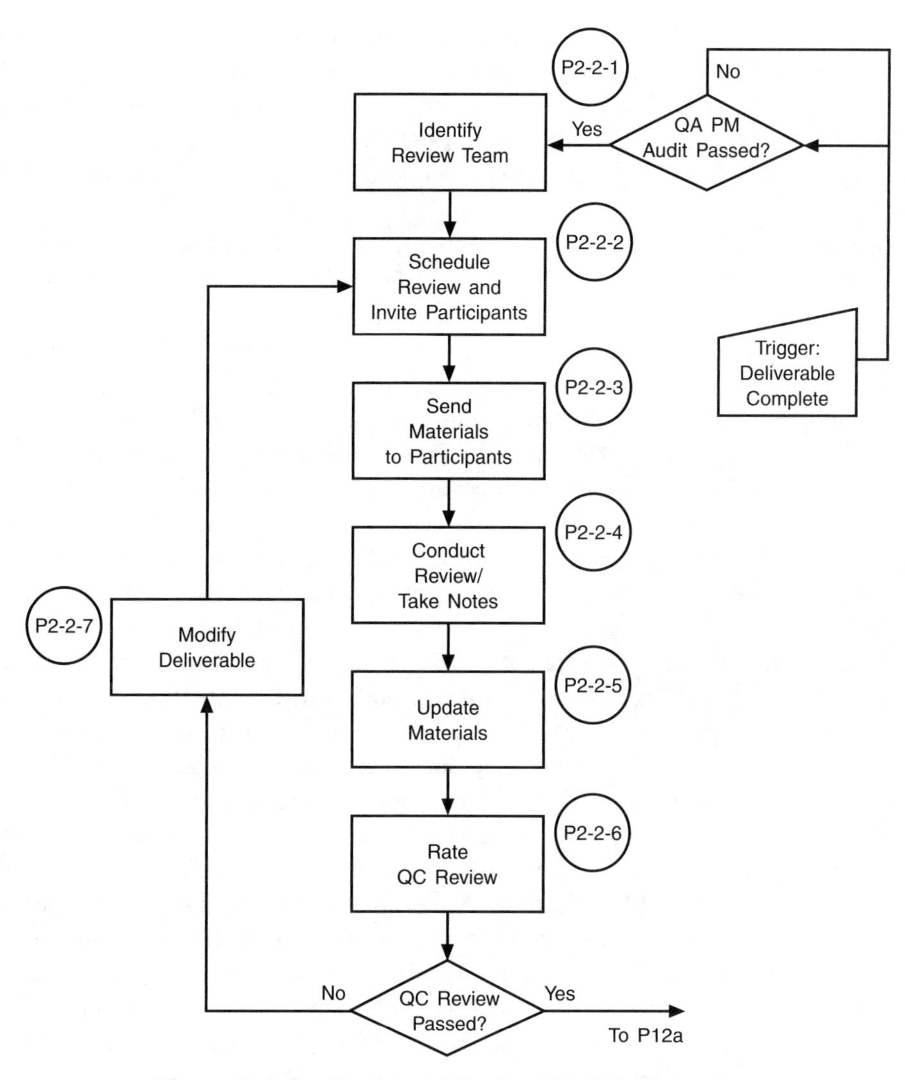

Figure 14.4 Quality Control Review (P2-2) Process Flow

A QC review is triggered when a deliverable is complete and the preceding QA project management audit is passed. The QC Review (P2-2) process runs in accordance with the flow in Figure 14.4.

The process is triggered by completion of a new deliverable document. The prerequisite for a QC review of a deliverable is the QA project management audit, which confirms that the deliverable has been produced in compliance with the project management process requirements.

Therefore, if the answer to the control point question (QA project management audit completed?) is no, the flow will return to the same control point and continue waiting for completion of the project management audit. Once the answer is yes, the process flow enters the Identify Review Team (P2-2-1) process.

This review should include the QA analyst, the project manager and the relevant delivery team members. Some QC reviews may require the presence of a pricing specialist or even a legal adviser. It is the project manager's and the QA analyst's joint responsibility to identify all relevant people who may contribute to the QC review. The objectives of the review are to document possible defects, omissions and unnecessary elements in the deliverable.

Once all the review team members are identified, they should be listed in a format similar to Table 14.15. The next process is Schedule Review and Invite Participants (P2-2-2), in which all the relevant people are invited to the review. The review must be scheduled at least seven days in advance, a conference room (or virtual equivalent) booked for the review and meeting invitations and the agenda sent to participants. Any participant who declines the invitation must ensure that a qualified representative is assigned to attend the review. The manager of the participant who declined the invitation must be notified. Without proper representation, the review must be rescheduled.

Participants will receive materials for review in the process Send Materials to Participants (P2-2-3). All materials related to the review must be sent to the participants by electronic or regular mail and received at least one week in advance. Materials include the agenda, documents, design and other relevant documentation.

At the designated time and place, the review is held in accordance with the process Conduct Review/Take Notes (P2-2-4), as described below. Comments and critiques are recorded following presentation of the deliverable to the review team. These comments usually indicate defects in the deliverable document. Each deliverable should have its own specific review method developed by the subject matter experts. Deliverable documents are often reviewed page by page. Defects or omissions are documented. Peer review of the architecture and design usually starts with a presentation by the author of the design and is followed by questions. Notes are taken by the project manager, QA analyst or a designated individual who understands the subjects discussed and is able to document any defects discovered and the required modification in a professional way. The person who takes notes normally should not lead the review or actively participate in the discussion, in order to have ample time (and focus) to concentrate on documenting all details of the review.

When all required modifications and fixes have been identified, the project manager must obtain confirmation from the review team by sending the notes with the requested modifications and fixes to all review team members. This will ensure that the review

Table 14.15 Project Team Members List

#	Name	Role	Area of Expertise	Phone	e-mail
1	A. Thomson	Quality Analyst	Quality Management	Ext. 2000	
2	K. Reid	System Architect	System Architecture	Ext. 2012	
3	M. Ortiz	System Analyst	System Analysis	Ext. 1812	
4	F. Norris	Project Manager	Project Management	Ext. 1948	
5	P. Ricer	Price Analyst	Pricing	Ext. 4290	

notes are correctly recorded and that only the required modifications and fixes will be made. The review activity should be documented in the project control book, including the date, attendees, agenda, results of the review and required modifications.

After the review has been held, defects will be removed and materials updated in the process Update Materials (P2-2-5), according to the notes taken during the review. The author of the deliverable must make all of the required updates and send the completed deliverable to the project manager and the QA analyst. Both must verify that no additional issues are left open.

The updated deliverable will be rated pass or fail by the QA analyst in the process Rate QC Review (P2-2-6). The QC review will be rated after the review materials have been updated and no issues are left open. Also, the QA project management audit, which immediately precedes this QC review, must have a pass grade. If this is indeed the case, the QC review rating is pass. If all QC issues have not been resolved, the review must be repeated. The QA analyst must present a full list of all remaining issues to the project manager for any follow-up modifications. The QA analyst must forward the QC review results to the project manager and the delivery manager.

If the QC review rating is pass, then the answer to the control point question (QC review passed?) will be yes, and the QC review is complete. If the answer is no, the delivery team will continue working on the deliverable until all issues are closed in the process Modify Deliverable (P2-2-7), and the process flow returns to process P2-2-2 to repeat the entire process again.

Quality Management Plan

The quality management plan is the guidance for "building in quality" through implementation of QA audits, QC reviews, quality reporting and the schedule of implementation. Much of the quality management plan consists of the quality approach and QA/QC activities built into the existing project plan. The quality management plan includes the following topics:

- Quality objectives
- QA and QC processes
- Established schedule and frequency of QA audits
- Schedule of QC reviews, tied to project deliverables
- Quality management roles and responsibilities
- Quality checklists
- Reporting quality issues

It may not be necessary to have a quality management plan as a separate document, because all the topics that would be covered have already been laid out in the quality management processes and checklists and documented in the quality section of the project control book. The schedule of specific QA and QC tasks should have been embedded in the existing project plan. The quality status must be reported to senior delivery management (on a monthly basis at a minimum) in the quality section of the project status report. Quality management responsibilities were discussed in previous chapters and must be documented in the project control book.

Quality Management Metrics

The following quality metrics must be collected and documented:

1. Number of scheduled QA audits in the project plan
2. Number of QA audits conducted from the list of scheduled audits
3. Number of unplanned emergency QA audits
4. Number of planned QA audits passed the first time
5. Number of planned QA audits passed the second time
6. Number of planned QA audits escalated
7. Number of emergency QA audits passed the first time
8. Number of emergency QA audits passed the second time
9. Number of emergency QA audits escalated
10. Number of deliverables identified in the project plan
11. Number of completed deliverables
12. Number of QC reviews
13. Number of QC reviews passed
14. Number of QC reviews failed

Section IV.
Construction/Tracking Frame

Construction/Tracking Frame

Purpose

The purpose of the Construction Frame is to:

- Produce project deliverables
- Monitor project progress and track implementation of all project activities against the project plan
- Implement and track scope changes
- Resolve and track issues
- Identify schedule and cost variances between the project plan and actuals
- Take corrective measures, when necessary, to bring the project back on track
- Ensure that all the deliverables are producing results that satisfy quality standards
- Ensure that all project management processes and quality standards are followed
- Keep records of all project events for quality management activities in order to avoid similar issues in future projects
- Gather statistics for improvement of the project management process

Roles and Responsibilities

The following roles and responsibilities will be required in the Construction Frame:

- **Project manager**—Responsible for ensuring that all steps in the project management process are followed and that the project team is involved in all stages of project implementation and tracking. The project manager tracks the construction and detailed design and reviews the status of the frame on a weekly basis with the client and on a monthly basis with senior business and senior delivery management. In some environments, the project manager reports on the status of the project to the senior business manager on a weekly rather than monthly basis.
- **Delivery team**—The group of project team members. The delivery team members will be responsible for producing the detailed design and the design implementation. They will also participate in quality audits and reviews. Each member of the delivery team will provide individual weekly status reports to the project manager during the scheduled project status meeting.
- **Client**—This is the lead client and a single focal point for the business. The client should participate in all status review meetings scheduled by the project manager. The client sometimes is referred to as the business project manager or the business area lead. The client usually reports to the senior business manager.
- **Technical lead**—Leading technical specialist and a member of the delivery team. The technical lead is responsible for producing the detailed design and the design implementation. The technical lead will assist the project manager in updating the work breakdown structure and gathering estimates from members of the delivery team.
- **Senior business manager**—Reviews monthly status reports with the project manager. The senior business manager usually reports to the project sponsor.
- **Senior delivery manager**—The person who owns a delivery budget and whose signature is required on the statement of work.
- **Project sponsor**—Major stakeholder who is responsible for the business success of the project, specifically, ensuring that the business objectives for which the project has been undertaken are met. The project sponsor is the owner of the overall project.
- **Quality assurance analyst**—Responsible for running quality management reviews and audits throughout the project life cycle in accordance with the quality management process. The quality assurance analyst runs all reviews in strict compliance with the quality management process and is responsible for documenting reviews, rating them and for follow-up when needed. The individual in this position should be someone with quality assurance training who is not directly associated with the project. However, for smaller projects, the project manager may play this role.

Inputs/Outputs

The Construction Frame processes interact with the rest of the project frames via entry and exit points, as shown in the Construction Frame process flow diagram (Figure 15.1):

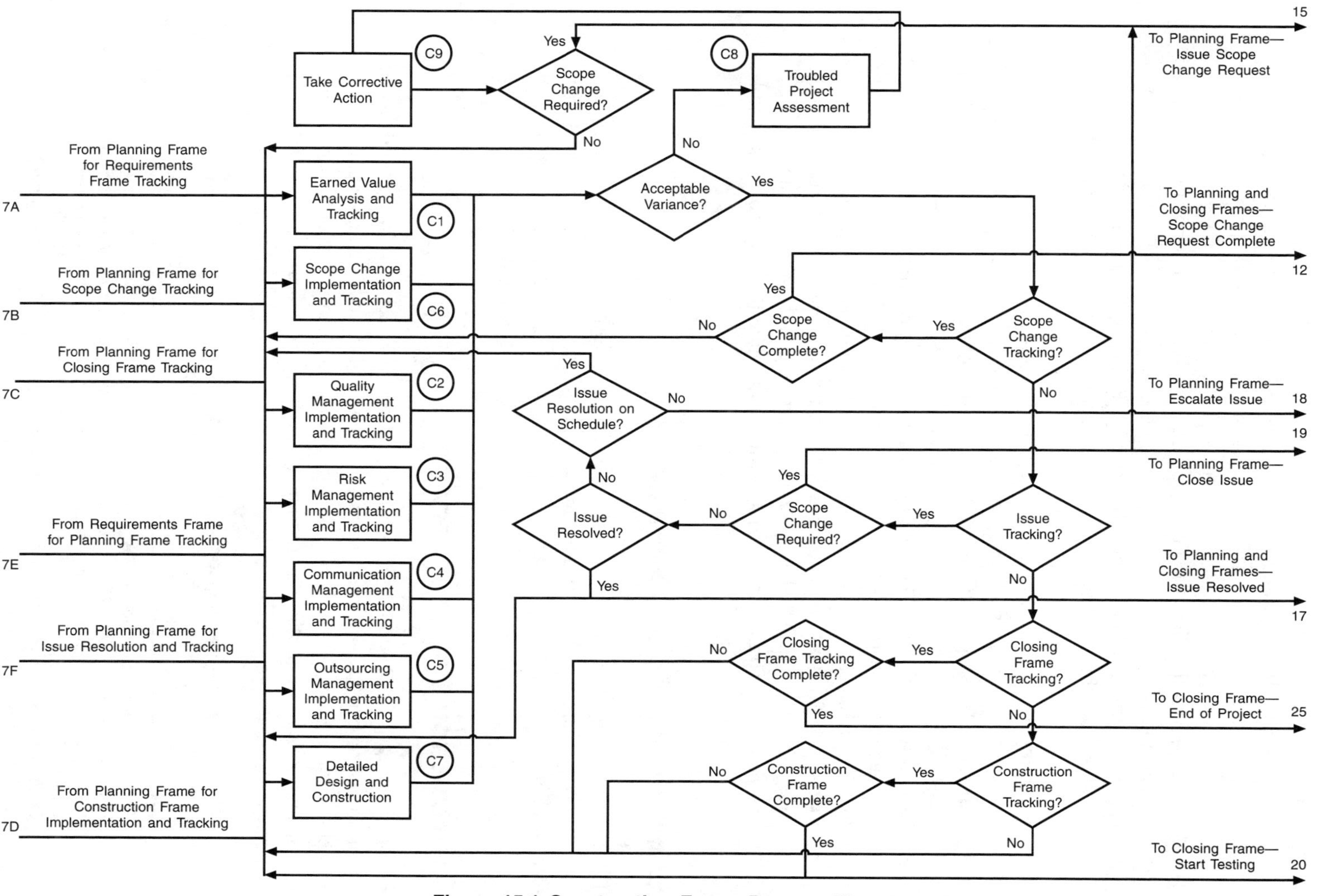

Figure 15.1 Construction Frame Process Flow

- **Entry point 7—**
  - ◇ 7A—From the Planning Frame for Requirements Frame tracking
  - ◇ 7B—From the Planning Frame for scope change tracking
  - ◇ 7C—From the Planning Frame for Closing Frame tracking
  - ◇ 7D—From the Planning Frame for Construction Frame implementation and tracking
  - ◇ 7E—From the Requirements Frame for Planning Frame tracking
  - ◇ 7F—From the Planning Frame for issue resolution implementation and tracking

- **Exit point 12**—To the Planning Frame and Closing Frame with notification that a scope change request is complete
- **Exit point 15**—To the Planning Frame with a request to issue a scope change request
- **Exit point 17**— To the Planning Frame and Closing Frame with notification that an issue has been resolved
- **Exit point 18**—To the Planning Frame for escalation of an issue
- **Exit point 19**—To the Planning Frame to close an issue for any reason
- **Exit point 20**—To the Closing Frame for testing and project closure
- **Exit point 25**—To the Closing Frame with notification that the project is complete

Construction Frame Requests

There are six business requests coming to the Construction Frame. As shown in Table 15.1, tracking of all activities in all frames, scope changes and issues is done in the Construction Frame.

Construction Frame Overview

In tracking and monitoring activities, the actual cost, project timeline and quality are continuously reported, comparing them against the plan and the acceptance criteria. Tracking entails obtaining facts and comparing them to planned values. Monitoring is the analysis to determine the required corrective measures, if any.

Table 15.1 Construction Frame Incoming Requests

Request	Coming from
Requirements Frame tracking (7A)	Planning Frame
Planning Frame tracking (7E)	Requirements Frame
Construction Frame implementation and tracking (7D)	Planning Frame
Closing Frame tracking (7C)	Planning Frame
Scope change implementation and tracking (7B)	Planning Frame
Issue resolution and tracking (7F)	Planning Frame

Actual size, effort, cost and schedule will be tracked against the baseline to measure project performance in accordance with the project tracking process, as described below. Baseline estimates and actual measurements are captured in the project control book. Estimates may be reviewed, but not necessarily changed, as part of project tracking, based on the planned versus actual variances. Corrective action means addressing those variances. The Construction Frame consists of the following processes:

1. Earned Value Analysis and Tracking (C1)
2. Quality Management Implementation and Tracking (C2)
3. Risk Management Implementation and Tracking (C3)
4. Communication Management Implementation and Tracking (C4)
5. Outsourcing Management Implementation and Tracking (C5)
6. Scope Change Implementation and Tracking (C6)
7. Detailed Design and Construction (C7)
8. Troubled Project Assessment (C8)
9. Take Corrective Action (C9)

Processes C2 through C6 were described in detail in the corresponding chapters in Section III on the Planning Frame, even though only the planning part of them is actually used in the Planning Frame. For clarity, it was necessary to provide there a complete description of the processes, since describing only the planning part would sacrifice the integrity of the process information. Therefore, at this point it would be beneficial for the reader to go back and review those chapters in Section III.

Implementation and tracking are part of the Construction Frame. Tracking is comparison of actual results with the expected values, as described below. Any discrepancy between them is evaluated against the acceptance criteria.

1. Implementation and tracking of earned value (process C1) means running the weekly earned value analysis (EVA) and then comparing actual performance results to the plan. This is the main method to determine whether or not the project is on track. EVA applies only to tasks included in the project plan, which also covers implementation of scope changes.
2. Implementation and tracking of quality management (process C2) includes conducting quality reviews and comparing the results with the expected quality. If, for example, an answer to one of the checklist questions is no, instead of the expected yes, there is a gap in the quality implementation, which must be fixed in order to get a passing quality audit grade.
3. Implementation and tracking of risk management (process C3) includes the scheduled risk assessments, risk evaluation and comparing actual risks that occurred and their remediation efforts to planned. When a risk occurs, an issue is automatically generated, which may or may not require initiating a scope change request for issue resolution.
4. Implementation and tracking of communication management (process C4) includes producing all planned reports and conducting all project status meetings,

as well as comparison between planned and actual communication. An important part of the process is actively managing stakeholders, as described in Chapter 9.

5. Implementation and tracking of outsourcing management (process C5) is the control and evaluation of subcontractors. It includes receiving status reports from subcontractors and comparing them to the subcontractor planned efforts. Issues may be generated when necessary.
6. Implementation and tracking of issue management generates issues in the Planning Frame when required, tracks issue resolution and generates scope change requests when necessary. When an issue is not resolved on schedule, it is escalated.
7. The Detailed Design and Construction (C7) process produces detailed design and construction based on the design. Both design and construction are main parts of the project plan and therefore are tracked.
8. Scope change implementation (process C6) produces the necessary scope change request design and implementation. Both are tracked against the project plan.

Processes C1 through C5 are executed in planned cycles, rather than continuously. For example, the EVA cycle most often is triggered and executed as a task once a week, before producing the weekly status reports. When it is done, it waits for the next week's cycle. The quality management process runs quality reviews periodically, as scheduled. When a review is completed, the process waits for a new cycle, which is triggered by the schedule.

Issue management stands on its own, since this is an event-driven process that takes place only when an issue comes up. Issue tracking is required only when an issue is open. As soon as a new issue comes up, the request comes through entry point 7F to start the issue resolution implementation and tracking cycle. Issue implementation is not a part of the overall project schedule. Issue resolution is the issue owner's responsibility, who may be a member of the delivery or business team. Issue status is tracked to determine whether the issue is expected to be resolved on time. The cost of issues is included in the cost of project overhead, unless the issue requires a scope change request. The Issue Management Planning (P4) process is described in detail in Chapter 8, even though issue tracking is done in the Construction Frame. An issue does not trigger the EVA process and may or may not trigger processes C2 through C5 unless it transforms into a scope change request.

Similar to issue tracking, scope change tracking is required only when a scope change is being implemented. It is no different than any other type of tracking and includes execution of processes C1 through C5 at their scheduled times. When scope change implementation is complete, the tracking process flow loops back to the beginning of the Construction Frame, waiting for any new scope change. When the scope change weekly tracking cycle is complete, but the scope change implementation is still in progress, the flow also loops back to the beginning, waiting for the new scheduled tracking cycle of the existing scope change. Other types of implementation tracking will usually continue in parallel with scope change implementation.

The exception to the cyclic execution is the Detailed Design and Construction (C7) process. This process is executed continuously until finished in accordance with the implementation schedule of tasks, but the EVA is tracked on a weekly based cycle.

When a request enters one of entry points 7A through 7F, it enables processes C1 through C7 to start their scheduled execution cycle in accordance with the schedule established in the Planning Frame. Tracking cycles do not necessarily start running all at the same time, but rather each one is enabled to start execution at the scheduled time.

Generally, the tracking process flow is multithreaded, and more than one type of tracking may run at the same time. For example, along with tracking project planning, it may be necessary at the same time to track scope change implementation and something else, subject to their tracking schedules. Most often, a frame implementation and scope change request tracking cycles are triggered once a week under normal circumstances, but the frequency may increase in the case of a troubled project. The issue tracking frequency depends on the gravity of the issue and may be as frequent as daily or as infrequent as weekly. If the resolution of a specific issue is temporarily put on hold, no tracking of that issue will be necessary until the issue resolution is resumed in accordance with the new schedule. The risk tracking frequency is once every two or three months and also before the beginning of a new frame or when a new risk is identified. When a planned or unplanned risk occurs, an issue is automatically generated, followed by a possible scope change as a consequence. Quality tracking occurs in accordance with the quality review schedule. While the project management quality audit may occur once every two or three months, some quality control reviews occur as often as the design or implementation of a new module is complete, which may be as frequent as several times a week.

Construction Frame Process Flow

The Construction Frame process flow diagram is shown on Figure 15.1. The Construction Frame starts when one or more inputs 7A through 7D activate one or more processes C1 through C7. Those processes will not be executed all at once; rather, each one will wait for its scheduled execution. When the execution starts, tracking results are obtained and compared to the planned values, and the process flow is directed to a control point question (acceptable variance?). If the variance between actuals and the planned values is unacceptable, as described earlier, and there is little chance of improving it in the normal course of the project within two or three weeks, the flow goes to the Troubled Project Assessment (C8) process to establish the reason for poor project performance and produce the corrective action plan. This plan is executed in the Take Corrective Action (C9) process, which in many cases leads to a project scope change. For example, if EVA shows that the cost or schedule has slipped over 10% and it is impossible to deliver the project on time at the approved budget, one of several options may be to reduce the scope of the project. This situation may happen when it is critical for the project sponsor to limit project expenditures to the approved budget or deliver the project by the established deadline, as explained earlier in Section III. In that case, the answer to the control point question (scope change required?) is yes, and the process flow is returned to the Planning Frame via exit point 15 to issue a new scope change request.

Sometimes it is possible to bring the project back on track by administrative actions without issuing a scope change request. For example, suppose the results of a quality assurance audit are poor, but no serious project damage has yet occurred. This does not require a scope change request, but the project *does* need corrective action to avoid potential issues. In other cases, the variance can be reduced and the project may get back on track by improving productivity and the work efficiency of the delivery team. In those cases, the answer to the control point question (scope change required?) is no, and the process flow goes back to the beginning of the tracking process, waiting for the next cycle of tracking and for the scheduled execution of one or more entry processes.

If the variance is acceptable, the project is on course, and the answer to the control point question (acceptable variance?) is yes. The next control point question (scope change tracking?) in the process flow is reached. From this point on, the process flow will depend on which of the following is being tracked:

1. Scope change implementation
2. Issue resolution
3. Closing Frame
4. Construction Frame implementation
5. Requirements Frame
6. Planning Frame

It was mentioned in earlier chapters that:

- Implementation of the Requirements Frame is done in the Requirements Frame
- Implementation of the Planning Frame is done in the Planning Frame
- Implementation of the Closing Frame is done in the Closing Frame
- Implementation of the Construction Frame is done in the Construction Frame
- Implementation of change requests is done in the Construction Frame

Tracking of all frames and change requests is done in the Construction Frame by using the EVA process (C1). Tracking is done in the scheduled cycles, at least once per week.

The EVA process does not make any distinction between different elements of tracking. When scheduled, it calculates variances between actual and planned values. The implementation of any frame and the tracking of the same frame are done simultaneously, and the completion of the frame implementation does not have any bearing on the tracking cycle. As long as there is something to track, the EVA cycle will be scheduled. EVA tracking will stop when the project is complete. Upon execution of one tracking cycle, the flow always returns to the beginning of the tracking process, waiting for the next scheduled tracking cycle. The only tracking situation in which EVA is not used is issue tracking, because *issues are not part of the project schedule*, unless they produce a scope change.

If the answer to the control point question (acceptable variance?) is yes, then the Construction Frame flows will be controlled and directed by a set of control point questions:

1. Scope change tracking?
2. Issue tracking?
3. Closing Frame tracking?
4. Construction Frame tracking?

If the scope change is being tracked, the answer to question #1 is yes. The implementation of the scope change is done in the scope change implementation process (C6), where the scope change design and implementation take place. Design and implementation is a continuous process, which runs according to the schedule, until completed. While the implementation continues, the process flow reaches a control point question (scope change complete?). If the scope change implementation is not yet finished, the answer is no, and the process flow loops back to the beginning of entry processes C1 through C6 for the next scheduled cycle of tracking. If the scope change is complete, the answer is yes, and the flow goes back via exit point 12 to the Planning Frame with an indication that the scope change request is complete.

If the scope change is not tracked in that particular cycle, then the answer to the control point question (scope change tracking?) is no, and the process flow reaches the next control point question (issue tracking?).

If the issue is being tracked, the answer to question #2 is yes, followed by the next control point question (scope change required?). If the issue cannot be resolved without changing the project scope, the answer is yes. That means the issue must be closed and a new scope change request must be opened. The request to close the issue is sent via exit point 19 and a request to open a scope change request is sent to the Planning Frame via exit point 15.

If a scope change is not required, the answer to the control point question (scope change required?) is no. The next control point question (issue resolved?) is reached. If the issue is resolved by this point, the answer is yes, and the indication is sent to the Planning Frame via exit point 17. At the same time, the process flow loops back to entry processes C1 through C6, waiting for the next scheduled tracking cycle.

If the issue is not resolved, the answer to the control point question (issue resolved?) is no, and the next control point question (issue resolution on schedule?) is reached. As long as there is no unacceptable delay in the issue resolution, the answer is yes, and the flow loops back to entry processes C1 through C6, waiting for the next tracking cycle. If there are delays in the issue resolution and there is no breakthrough in sight, the answer is no, and the request to escalate goes to the Planning Frame via exit point 18.

If no issue is being tracked in this tracking cycle, the answer to the control point question (issue tracking?) is no. The next control point question (Closing Frame tracking?) is reached. If the current tracking cycle tracks the Closing Frame, the answer is yes, and the next control point question (Closing Frame tracking complete?) is reached. If the answer is yes, the entire project is complete, and thus no more tracking is required. The request to close the project is sent to the Closing Frame via exit point 25. If the answer is no, project execution continues and the process flow loops back to the beginning of the process.

If the answer to the control point question (Closing Frame tracking?) is no, the next control point question (Construction Frame tracking?) is reached. If the Construction

Frame is not being tracked in this tracking cycle, the answer is no, and the process flow loops back to the beginning of the Closing Frame. If the Construction Frame is being tracked in this cycle, the next control point question (Construction Frame complete?) is reached. If the answer is no, the flow loops back to the beginning of the Closing Frame, waiting for the next tracking cycle. If the Construction Frame is complete, the process flows to the Closing Frame via exit point 20 to start testing and Closing Frame implementation. At the same time, the flow loops back to the start of the cycle, waiting for tracking of the Closing Frame or a scope change or issue, whichever is next in the schedule.

If the answer to all of the above control point questions is no, this means that either the Requirements Frame or the Planning Frame is being tracked. The process flow will return to input processes C1 through C6, waiting for the next scheduled tracking cycle. When tracking of those processes takes place, no special notification is required, provided the variance is acceptable. The process flow will return to input processes C1 through C6, waiting for the next scheduled tracking cycle.

Earned Value Analysis and Tracking (C1)

Earned Value Analysis Methodology

EVA is a widely used method to keep track of project work and report project financials and project health. EVA calculates the relationship between the actual cost of work, the planned cost of work and the planned cost of the work actually completed. The method avoids the problem of reporting progress in nontransparent ways, such as an outsourcing vendor indicating that the project is "half done" simply because four out of eight weeks of the schedule has elapsed.

The EVA method can also help predict how much the project will cost and when it will end. EVA is supported by most project scheduling and tracking tools. Let us say that project X has five tasks and an estimated duration of six months. The project start date is January 1. The planned cost of the project, called *budget at completion* (BAC), is $100,000, as shown in Table 15.2. Let us also assume that today is March 31. The actual spending so far is $60,000. Is this good or bad? The answer will depend on how much work has been completed ("earned") for $60,000, which is 60% of the budget. If only 30% of the scheduled work is complete, this is bad. On the other hand, if 70% of the work is complete, this is good.

Table 15.2 Budget at Completion Calculation Example

Project X Tasks	Planned Cost
Task 1 (January)	$20,000
Task 2 (February)	$20,000
Task 3 (March)	$20,000
Task 4 (April)	$20,000
Task 5 (May)	$20,000
BAC	**$100,000**

EVA is a method that takes all of this into consideration and puts the answers in monetary terms. The following are major elements of EVA, which are usually generated by project scheduling tools, provided the schedule is up to date and resource rates are correct.

Planned value (PV) is the planned cost of work performed at the time EVA is calculated. In our case, PV on March 31 is $60,000.

Earned value (EV) is the planned cost of work actually completed. If by March 31 we finished task 1, task 2 and 50% of task 3, the budgeted cost of work performed is $50,000, instead of $60,000 worth of work planned. In this case, we are behind schedule. However, if we finished task 1, task 2, task 3 and 50% of task 4, EV is $70,000, which makes the project ahead of schedule.

Actual cost (AC) is the accrued cost of the work completed. If by March 31 we finished task 1, task 2 and 50% of task 3 at AC = $65,000, we spent $15,000 more than the planned cost of the work performed ($50,000) or, in other words, we are $15,000 over budget.

Based on the cost and schedule for the above tasks, on March 31 PV = $60,000. Let's look at five variations of EV and AC and try to determine the health of project X in terms of cost and schedule, as shown in the example in Table 15.3.

- **Case 1**—By the end of March, the project team accomplished work worth $60,000 (EV), which is exactly what was planned. This means that the project is exactly on schedule. AC = $60,000, which means the project cost is exactly as budgeted.
- **Case 2**—EV = $70,000, which means that the work performed exceeds the planned worth of the work ($60,000). This means that the project team accomplished more work than planned and the project is ahead of schedule. AC = $70,000, which means that the actual cost of the work performed is the same as the budgeted cost of the work performed. The conclusion is that the project is within budget.
- **Case 3**—EV = $60,000, which is exactly the same as the planned work. The project is on schedule. AC = $70,000, which is $10,000 over the planned cost of the work performed. The project is $10,000 over budget.
- **Case 4**—EV = $50,000, which means the project is behind schedule. The team was supposed to complete work worth $60,000 by the end of March, but only completed work worth $50,000. AC = $40,000. Since the work performed is worth $50,000 but the cost is only $40,000, the project is below budget.
- **Case 5**—EV = $70,000. The project is ahead of schedule. AC = $60,000. The project is below budget.

Table 15.3 Earned Value Analysis Inputs

	PV	EV	AC
Case 1	$60,000	$60,000	$60,000
Case 2	$60,000	$70,000	$70,000
Case 3	$60,000	$60,000	$70,000
Case 4	$60,000	$50,000	$40,000
Case 5	$60,000	$70,000	$60,000

NOTE When a project in real life is consistently ahead of schedule and below budget, this does not always mean excellent team performance, but rather may mean the project was not planned properly and was poorly estimated, which is another excuse for senior managers to use the project manager as a piñata. Exceeding project performance always arouses doubts about the project manager's planning skill.

EVA uses BAC, PV, EV and AC to calculate the following values (an example of EVA calculation is provided in Table 15.4).

Cost performance index (CPI) shows the cost efficiency of the project team in producing the required work:

$$CPI = EV/AC$$

If CPI is less than 1, the actual cost exceeds the budgeted cost. Correspondingly, if CPI is greater than 1, the actual cost is below the budgeted cost.

Cost variance (CV) is the difference between the budgeted and the actual cost of the work produced in dollars or other currency:

$$CV = EV - AC$$

If CV is negative, the work performed is worth less than the actual cost, which means the project is over budget. If CV is positive, the work performed is worth more than the actual cost.

Schedule performance index (SPI) shows the efficiency of the project team in producing the required work:

$$SPI = EV/PV$$

If SPI is less than 1, then producing the work takes longer than scheduled, which means the project is behind schedule. If SPI is greater than 1, then producing the planned work took less time than scheduled.

Schedule variance (SV) is the difference between the budgeted cost of the work produced and the budgeted cost of the planned work in dollars or other currency:

$$SV = EV - PV$$

Table 15.4 Example of Earned Value Analysis Calculation

	PV	EV	AC	CV	CPI	SV	SPI
Case 1	$60,000	$60,000	$60,000	$0.0	1.0	$0	1.0
Case 2	$60,000	$70,000	$70,000	$0.0	1.0	$10,000	1.17
Case 3	$60,000	$60,000	$70,000	–$10,000	0.86	$0	1.0
Case 4	$60,000	$50,000	$40,000	$10,000	1.25	–$10,000	0.83
Case 5	$60,000	$70,000	$60,000	$10,000	1.17	$10,000	1.17

If SV is negative, less work than planned was performed, which means the project is behind schedule. If SV is positive, more work than planned was performed.

Estimate at completion (EAC) is the estimated cost of the project by the time it is completed, provided the CPI stays just about the same:

$$EAC = BAC/CPI$$

Variance at completion (VAC) is the difference between the planned budget and the estimated cost at completion:

$$VAC = BAC - EAC$$

Estimated completion date (ECD) is the date when the project is expected to be completed assuming the SPI stays about the same. The calculation is cumbersome for manual calculation, but it is easy to do using the following formula in a spreadsheet:

WORKDAY (Start_Date,(NETWORKDAYS(Start_Date, Planned_End_Date)/SPI))

CPI and SPI allow us to predict the cost and the project end date based on past performance. If CPI is 0.8, the project will cost 20% more or $120,000. If CPI is 1.2, the project will cost 20% less or $80,000. If SPI is 0.8, it will take 20% longer or six months to complete the project instead of the planned five months. If SPI is 1.8, it will take 20% less time or four months to complete the project.

EAC and VAC provide new project cost estimates and the variance between EAC and BAC. ECD provides the estimated project completion date, based on the SPI.

An EVA tool automates calculation and provides unambiguous information on project health. Table 15.5 (A to C) shows three sample calculations an EVA tool can generate for different EVA values.

Most project scheduling tools support EVA. In order to be able to use a scheduling tool, the following must be done when developing the project schedule:

1. Assign all tasks in the project plan to specific resources.
2. Enter task dependencies.
3. Level the schedule, which distributes the workload to prevent overloading resources.
4. Enter the rate of each resource.
5. Baseline the plan.
6. Start receiving weekly status reports, as discussed below.
7. Use a *tracking Gantt chart* to update the project plan weekly.

The scheduling tool will provide BAC, PV, EV and AC.

It is very important to get weekly project status information from team members (the report template was provided in Table 9.2), outlining the status of every noncompleted task due or task being performed, and enter the actuals in a tracking Gantt chart. For example, if today is July 7, the report in Table 15.6 shows the tracking information that must be entered into a scheduling tool in order to calculate EVA.

Table 15.5A Earned Value Analysis Sample 1

EVA Date: 12/15/2013		
Budget at completion (BAC)—initial budget	$100,000	
Project start date (mm/dd/yyyy)	10/01/2013	
Planned end date (mm/dd/yyyy)	03/01/2014	
Actual cost of the work performed (AC)	$60,000	
Budgeted cost of the work performed (EV)	$40,000	
Budgeted cost of the work scheduled (PV)	$50,000	
Cost variance (CV = EV – AC)	–$20,000	Over budget
Schedule variance (SV = EV – PV)	–$10,000	Behind schedule
Cost performance index (CPI = EV/AC)	0.67	66.7 % cost performance
Schedule performance index (SPI = EV/PV)	0.80	80.0 % schedule performance
Estimate at completion (EAC = BAC/CPI)	$149,998	
Variance at completion (VAC = BAC – EAC)	–$49,998	
Estimated completion date	04/07/2014	

Table 15.5B Earned Value Analysis Sample 2

EVA Date: 12/15/2013		
Budget at completion (BAC)—initial budget	$100,000	
Project start date (mm/dd/yyyy)	10/01/2013	
Planned end date (mm/dd/yyyy)	03/01/2014	
Actual cost of the work performed (AC)	$52,000	
Budgeted cost of the work performed (EV)	$52,000	
Budgeted cost of the work scheduled (PV)	$50,000	
Cost variance (CV = EV – AC)	$0	Within budget
Schedule variance (SV = EV – PV)	$2,000	On schedule or ahead of schedule
Cost performance index (CPI = EV/AC)	1.00	100.0 % cost performance
Schedule performance index (SPI = EV/PV)	1.04	104.0 % schedule performance
Estimate at completion (EAC = BAC/CPI)	$99,999	
Variance at completion (VAC = BAC – EAC)	$1	
Estimated completion date	02/22/2014	

Table 15.5C Earned Value Analysis Sample 3

EVA Date: 12/15/2013		
Budget at completion (BAC)—initial budget	$100,000	
Project start date (mm/dd/yyyy)	10/01/2013	
Planned end date (mm/dd/yyyy)	03/01/2014	
Actual cost of the work performed (AC)	$48,000	
Budgeted cost of the work performed (EV)	$52,000	
Budgeted cost of the work scheduled (PV)	$50,000	
Cost variance (CV = EV – AC)	$4,000	On or below budget
Schedule variance (SV = EV – PV)	$2,000	On schedule or ahead of schedule
Cost performance index (CPI = EV/AC)	1.08	108.0 % cost performance
Schedule performance index (SPI = EV/PV)	1.04	104.0 % schedule performance
Estimate at completion (EAC = BAC/CPI)	$92,308	
Variance at completion (VAC = BAC – EAC)	$7,692	
Estimated completion date	02/24/2014	

Table 15.6 Weekly Team Status Report Sample

Weekly Team Status Report			
Project Name: ______		Project Manager: ______	
Team Member: ______		Week Ending: ______	
Task ID: 1		% Completed:	100%
Scheduled Start Date:	July 1	Actual Start Date:	July 2
Scheduled End Date:	July 3	Actual End Date:	July 5
Budgeted Hours:	16	Actual Hours:	18
Additional Time Needed:	No	Reason for Delay: Previous task was not completed on time; estimates were incorrect	
Task ID: 2		% Completed:	60%
Scheduled Start Date:	July 4	Actual Start Date:	July 6
Scheduled End Date:	July 9	Actual End Date:	______
Budgeted Hours:	24	Actual Hours:	16
Additional Time Needed:	No	Reason for Delay:	

When resource availability is 100%, it should be assumed that resources work on the tasks assigned to them for 6.5 hours a day. The rest of the day (1.5 hours) is not shown on the plan, because it is daily overhead, like reading e-mail, taking phone calls, attending meetings, etc. However, the cost of overhead must be taken into consideration when calculating EVA. One option is to add 20% to the effort for every task when estimating work in the Planning Frame. In this case, assuming an 8-hour workday, the task effort is 6.4 hours and overhead is 1.6 hours.

Earned Value Analysis Tracking

The EVA tracking process consists of the following steps:

1. Receive status reports from all delivery team members.
2. Enter the status data into the tracking chart of the scheduling tool.
3. Obtain BAC, PV, EV and AC from the scheduling tool.
4. Enter BAC, PV, EV and AC into the EVA tool (if a tool is used for EVA calculations).
5. Based on the information entered, the EVA tool will calculate CPI, CV, SPI, SV, EAC, VAC and ECD, or those values can be calculated manually using the formulas given above.

Earned Value Analysis Tracking Acceptance Criteria

EVA is used when tracking the implementation of:

1. Requirements Frame
2. Planning Frame
3. Construction Frame
4. Closing Frame
5. Scope change
6. Outsourced subcontract

EVA is not used when tracking issues. If a scope change request is generated during issue tracking, then the scope change is tracked.

During tracking cycles, EVA acceptance criteria must be used to determine whether the results are acceptable. While up to a 10% cumulative variance from the cost, schedule or calculated variance at completion is usually acceptable, a higher number also may be temporarily acceptable if it is reasonable to believe that there is a good chance of improving the variance.

Quality Management Implementation and Tracking (C2)

Quality management implementation consists of the planned quality reviews and audits. Quality management tracking compares the planned versus the actually executed quality reviews and audits. The quality audit and quality review process was discussed in Chapter 14 as part of the Quality Management Planning (P2) process.

Quality management audits and reviews take place in accordance with the schedule developed in the Planning Frame. Quality reviews are usually performed when project deliverables are being developed. Quality audits are usually performed every three to four months. If a serious project issue is found, they will be executed every one or two weeks until the quality issue is resolved and project management processes are strictly enforced.

Quality Management Tracking Acceptance Criteria

Quality management acceptance criteria are determined by the quality assurance audit or quality control review passing criteria. The results of a quality assurance audit are acceptable when a pass grade is issued, as discussed in Chapter 14. If all gaps have not been resolved, a flag is raised to indicate that the quality assurance audit results are unacceptable. During quality management tracking, if the answer to the control point question (acceptable variance?) is no, the flow is directed to the Troubled Project Assessment (C9) process.

The results of a quality control review are acceptable when a pass rating is issued, as discussed in Chapter 14. If after modification of the deliverable a pass rating is not issued, a flag is raised to indicate that quality control review results are unacceptable. During quality management tracking, if the answer to the control point question (acceptable variance?) is no, the flow is directed to the Troubled Project Assessment (C9) process.

Whether or not quality management results are acceptable is determined when the quality assurance or quality control process is complete and there is an opportunity to fix unacceptable results before admitting failure.

Risk Management Implementation and Tracking (C3)

The risk management process consists of three components:

1. Risk assessment
2. Risk planning
3. Risk management implementation and tracking

The first two were described in Chapter 6. Risk management implementation is the execution of planned and sometimes unplanned risk assessments. The planned and scheduled risk assessments run according to the plan developed in the Planning Frame. Unplanned risk assessments occur when a new risk is discovered.

Risk tracking is the process of monitoring risk occurrence. When a planned or unplanned risk occurs, it is initially treated as an issue in accordance with the issue management process, as discussed in Chapter 8. Some issues require generating a project scope change request. The risk containment plan is executed as a scope change in accordance with the Scope Change Control (P7) process (see Chapter 10). When a scope change is generated, then the project plan is adjusted for each risk that occurred and its implementation is tracked.

The following elements of risk management are documented:

- Each risk assessment and the reason for the assessment
- Total number of risk assessments
- All occurrences of previously identified risks (this comes from the issue management metrics)
- All occurrences of new previously unknown risks (this comes from the issue management metrics)
- Planned cost of each risk mitigation and the overall planned cost of all risk mitigations
- Actual cost of each risk mitigation and the overall actual cost of all risk mitigations
- Planned cost of each risk that occurred and the overall planned cost of risks (this comes from the issue management and scope change control metrics)
- Actual cost of each risk that occurred and the overall actual cost of risks that occurred (this comes from the issue management and scope change control metrics)

Risk Management Tracking Acceptance Criteria

Risk management tracking results are acceptable when the actual cost of risk management does not exceed 10% of the planned cost. This number comes from the scope change EVA, because risk mitigation is usually guided by scope change management, as discussed earlier. When an unknown risk occurs, the total cost of all unknown risks cannot exceed the margin allocated for unknown risks, which is usually 10% of the project cost.

Communication Management Implementation and Tracking (C4)

The purpose of communication management implementation is to maintain communication among project team members in the delivery, outsourcing and business organizations in order to generate and exchange project-related information and create understanding between the originator of information and the receiver. The types and methods of communication, as well as a description of proactive communication, meetings, forms, reports, the communication plan, orientation and training, were presented in Chapter 9. Communication management tracking compares actual dates when meetings, orientation and training took place versus the communication plan.

Communication Management Tracking Acceptance Criteria

Communication management tracking results are acceptable if no communication failure is detected. Signs of communication failure are two or more of the following occurrences in one month:

- Meetings are not scheduled, the agenda is not sent or participants are not invited in a timely fashion
- Meetings are not held in accordance with the project plan or attendance of key people in such meetings is limited
- Status reports are not produced, not communicated or not reviewed by the addressees
- Not all major stakeholders are identified or major stakeholders are not participating in the project

Outsourcing Management Implementation and Tracking (C5)

Outsourcing management planning was discussed in Chapter 12. Tracking of the outsourced parts of the project is based on status and financial reports provided by the outsourcing organization and is no different than implementation tracking. If variances are detected in quality, cost or schedule of deliverables, then an unacceptable variance flag is raised.

Scope Change Implementation and Tracking (C6)

Scope change implementation is the design and implementation of the scope change plan in order to deliver a scope change. Scope change tracking is the tracking of actuals versus the plan. Since the scope change plan is usually embedded in the overall project plan, tracking it is a part of the overall project implementation tracking. EVA may be calculated separately for scope change implementation using an EVA tool, even though it may be difficult to separate EVA of a scope change and EVA of a project implementation. The scope change process and flow diagram were presented in Chapter 10.

Scope Change Tracking Acceptance Criteria

Scope change tracking results are acceptable if:

- EVA calculation results for the scope change are acceptable (±10% of the planned scope costs and schedule)
- Quality management ratings for the scope change activities are acceptable

Detailed Design and Construction (C7)

Tasks required to implement detailed design and construction are different for different types of projects. Because planning of the Construction Frame, including detailed design, is done in the Planning Frame, the main goal of process C7 is implementation of the existing project plan.

Detailed Design and Construction Tracking Acceptance Criteria

Detailed design and construction tracking is acceptable if:

- EVA calculation results are acceptable (±10% of the planned costs and schedule)
- Quality management results are acceptable (acceptable rating on planned quality assurance audits and quality control reviews)

Troubled Project Assessment (C8)

The purpose of the Troubled Project Assessment (C8) process is to determine major reasons why the project is unable to proceed as planned. Most problems do not happen suddenly. Most projects start running into trouble soon after beginning. Some authors on this subject suggest that the main reason is not following the change request process; others blame lack of proper issue and risk management processes. Many blame the delivery team, sponsors and management for the project outcome.

All of them are correct, but there are many more reasons for poor project performance. The question is how to find those reasons and whether it is really possible and necessary to find all of them.

In 1906 the Italian economist V. Pareto observed that 20% of the people owned 80% of the wealth. Scientists proved that his observation, called the 20/80 or the Pareto principle, is correct for most business areas:

- Software—20% of the reported software bugs cause 80% of the errors and crashes.
- Service—20% of customers take up 80% of a service organization's resources.
- Health care—20% of patients use 80% of health care resources.
- Law—20% of criminals commit 80% of crimes.
- Engineering—20% of components are responsible for 80% of failures.

The Pareto principle may be generalized to claim that roughly 80% of the effects come from 20% of the causes. This makes the task of determining causes for poor project performance easier, since fixing only 20% of the causes will improve project performance by 80%.

The method to determine causes is called cause-and-effect analysis, which involves analyzing the situation and coming up with major reasons for poor project performance. The tool used for the analysis is the cause-and-effect diagram (Figure 15.2), also called the fishbone (for the way it looks) or Ishikawa (after its inventor, Dr. Kauro Ishikawa) diagram. Using this tool, it is possible to utilize ideas from all team members and consider all possible causes of the problem. This approach uses a brainstorming session with participation of all relevant team members and clients. It is a good idea to have an outside moderator run the session. Conclusions are presented to management and sponsors.

Cause-and-effect analysis consists of the following steps:

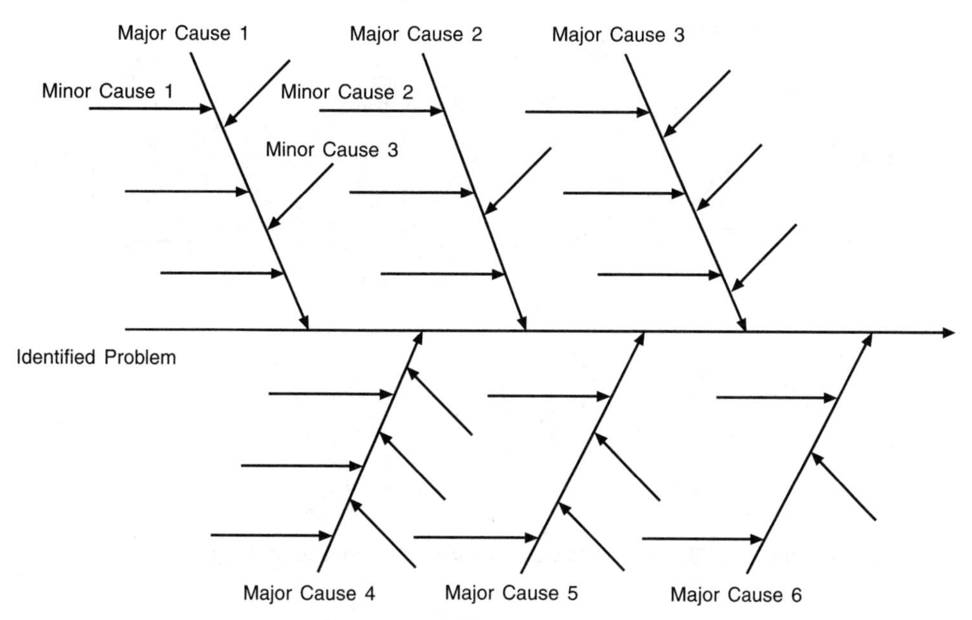

Figure 15.2 Fishbone Diagram

1. Identify the problem.
2. Identify the possible major causes involved (these form the "ribs" of the diagram).
3. Break down major causes into possible minor causes (done by asking "why" repeatedly).
4. Analyze the diagram and provide more details about all relevant minor causes. Additional investigation may take place.
5. Propose a solution to improve project performance.

A fishbone diagram for an information technology problem is shown in Figure 15.3. The branches required will depend on the type of problem. For example, if the project slipped the schedule during implementation, then the user training branch is not required. The database administration branch may be needed only if developers have issues with the database. The architecture branch may or may not be required, depending on the problem. On the other hand, a communication branch (missing in Figure 15.3) may be needed if the client has not clearly explained the requirements. Also, a management branch may be needed if the existing corporate environment is indifferent to violation of project management processes or if some managers ignore processes. If an issue occurs during acceptance testing or after the product is released to production, then the user training branch may be relevant.

When the list of possible reasons for poor project performance is analyzed, a meeting should be held to identify ways to improve each problem area identified and develop a corrective action plan. Team members and clients should participate in those efforts. Key team members have good ideas about reasons for failure, even though they are not always

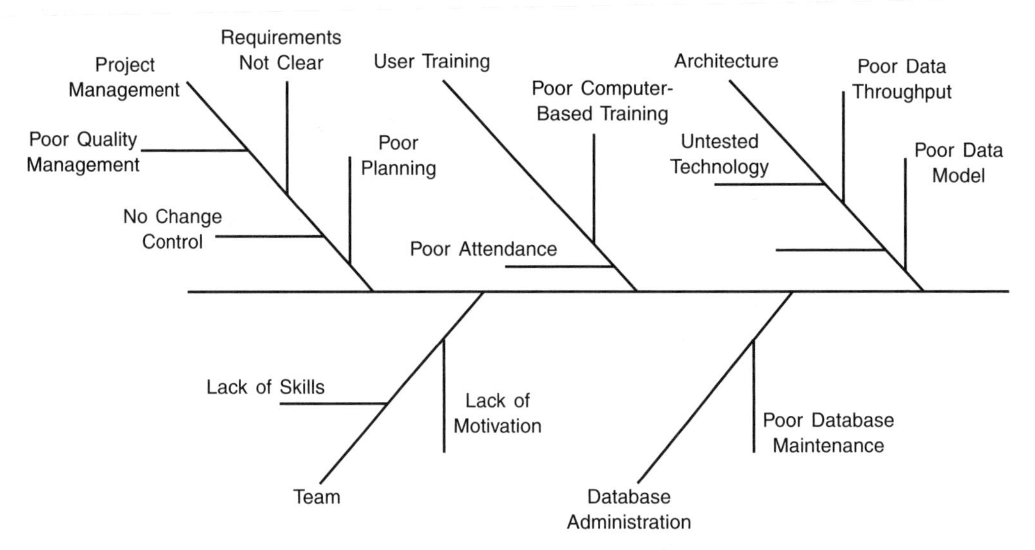

Figure 15.3 Fishbone Diagram for Information Technology Project

outspoken in public. The project manager should meet with them informally, one at a time, and ask for their advice, trying to understand their point of view. The corrective action plan must provide the schedule and deadline for each corrective action taken. It must specify by name those responsible for implementation of each corrective action.

It is important to ensure that team conflicts are not an issue. If team conflicts exist, the methods discussed in Chapter 9 may be used to resolve them. Changes to the team can be introduced to improve team morale. As a last resort, the poor performers should be removed. Reducing project scope, thus reducing complexity of the project, can be considered. If undue management interference in the project is the reason for frequent changes in the project plan, the project manager should make sure this opinion is shared by the majority of the team and then present this view to management. This will not win the project manager favor with management, but failure to deliver the project will be much worse.

The corrective action plan should not take longer than two or three weeks to implement. This is usually sufficient to improve project performance. If performance does not improve, either the analysis is wrong or recovery is not possible. Indeed, the recovery may not work even with a good recovery plan when the corporate environment does not encourage practicing good project management processes.

The following are a few samples of causes and effects of poor performance in many projects:

- If the project scope is not fully documented or scope changes are not properly processed, as described in Chapters 3, 4 and 10, there will be scope leak in the project and the final product will be different from what was documented. Clients may and will forget about their requests to change the scope. The project cost will overrun the budget if project modifications are not properly authorized and priced.

- If the estimating process presented in Chapter 11 is not followed, the project will end up with overly optimistic estimates.
- If risks are not managed as described in Chapter 6, the project may end up with many issues and change requests, which will eat up the project budget.
- If communication with team members, management and clients does not follow the process outlined in Chapter 9, the project will be uncontrollable.
- If the project is not planned properly or progress is not tracked meticulously, the project is destined for failure.
- If the quality review process discussed in Chapter 14 is not followed, the product developed will not fit the purpose for which it was designed.

It is very important that the project manager does not commit to a new baseline under pressure, but rather creates a new achievable, controllable and properly funded plan first.

Recovery is not always possible or even needed. The following are some examples of this situation:

- None of the required project skills will be available in the foreseeable future.
- The sponsor and the business lost interest in the project, because the business benefits are no longer apparent.
- Technology changed to such an extent that the product being developed will be outdated by the time it is completed.

Take Corrective Action (C9)

Corrective action is taken in order to eliminate reasons affecting project performance and improve it to be in line with the project baseline, quality or expected results. The corrective action plan describes when and what corrective action should be taken.

Often, the project manager does not have the authority to implement corrective action without management assistance, especially if some of the reasons for poor performance are due to client or management actions. Therefore, the developing organization management and the sponsor must approve the corrective action plan and participate in assignment of action owners.

Implementation of the corrective action plan is tracked daily until the implementation plan is completed and a new project baseline is established. Issues related to staff members who do not meet the deadlines in the corrective action plan will be immediately escalated to senior management.

Section V.
Closing/Testing Frame

Closing/Testing Frame

Purpose

The Closing Frame is the frame in which the project ends. Its purpose is to:

- Prepare the environment, develop a test plan and execute acceptance testing
- Develop training materials and conduct user training sessions
- Ensure that all project management processes and standards are followed
- Verify that all project deliverables are produced in accordance with the quality standards
- Keep records of all project events in order to avoid similar issues in future projects
- Roll out the project to production or deliver all project materials to the client
- Release project resources
- Conduct and document a lessons learned analysis
- Gather the Closing Frame statistics for project management process improvement

Roles and Responsibilities

The following roles and responsibilities will be required in the Closing Frame:

- **Project manager**—Responsible for ensuring that all steps of the project management process are adhered to. The project manager ensures that the delivery and client teams are involved in all stages of the Closing Frame. The project manager is responsible for the following frame activities:
  - ◇ Track the frame implementation

- ◇ Manage the acceptance test support team activities
- ◇ Allocate the project team resources for acceptance testing
- ◇ Close the project and reallocate all project resources
- ◇ Conduct the project lessons learned analysis
- ◇ Review the status on a weekly basis with the client and on a monthly basis with senior business and senior delivery management (in some environments, the project manager reports the status of the project to the senior business manager on a weekly rather than monthly basis)

- **Client**—This is the lead client and a single focal point for the business. The client must participate in all status review meetings scheduled by the project manager. The client is sometimes referred to as the business project manager or the business area lead. The client usually reports to the senior business manager. During acceptance testing, the client has the following responsibilities:
  - ◇ Overall responsibility for development of the acceptance test plan
  - ◇ Schedule acceptance testing with all relevant project participants
  - ◇ Obtain commitment of all resources needed for acceptance testing
  - ◇ Ensure that all steps of the testing procedure are followed and documented
  - ◇ Ensure that all test scripts are documented
  - ◇ Ensure that all defects are documented in the acceptance test incident recording tool
  - ◇ Approve the acceptance test completion document
  - ◇ Ensure that acceptance testing is completed according to the schedule, budget and quality standards
  - ◇ Resolve business issues that arise out of testing
  - ◇ Approve change requests that arise out of testing
- **Delivery team lead**—Responsible for coordinating acceptance test support team activities and allocating the project team resources to support the acceptance testing.
- **Test manager**—A member of the business team assigned by the lead client. The test manager has the following responsibilities:
  - ◇ Develop the acceptance test plan and ensure that it conforms to the test strategy document
  - ◇ Ensure that the test strategy document exists prior to acceptance testing
  - ◇ Ensure that detailed test conditions or test scripts are produced
  - ◇ Ensure that entrance and exit criteria are defined prior to acceptance testing
  - ◇ Review the acceptance test plan with the acceptance test team
  - ◇ Regularly review test progress with the acceptance test team
  - ◇ Ensure that root cause analysis is performed for all defects found
  - ◇ Ensure that completion criteria are met for all tests before the acceptance test completion document is signed off on
  - ◇ Manage test cycles
  - ◇ Manage the resolution of acceptance test team issues and queries
  - ◇ Ensure that all defects and problems are reported immediately and followed up until they are resolved

  - ◇ Participate in the daily defect review meetings with the acceptance test support team
- ◆ **Acceptance test team**—Members are assigned by the lead client from the pool of business users. The following responsibilities are assigned to them by the lead client:
  - ◇ Develop test cases and test scripts and have them approved by the test manager
  - ◇ Develop test schedule/calendar, based on the acceptance test plan, for the test cases assigned to them
  - ◇ Identify test data and ensure that data is available and accessible during acceptance testing
  - ◇ Execute test scripts
  - ◇ Produce defect reports
  - ◇ Rerun test cases after correcting defects
- ◆ **Acceptance test support team**—This team is different from the acceptance test team. Its members are selected from the delivery team pool. The following responsibilities are assigned to them by the project manager:
  - ◇ Provide timely support for the acceptance test team
  - ◇ Resolve defects
  - ◇ Perform root cause analysis for all defects
  - ◇ Participate in the daily defect review meetings with the acceptance test manager
  - ◇ Provide test infrastructure
  - ◇ Serve as supporting resources for running scheduled tests
  - ◇ Provide related test reports
  - ◇ Provide interfaces to other systems according to the acceptance test plan
- ◆ **Test coordinator**—The member of the delivery team who receives defect notifications, assigns resources from the test support team pool to fix defects, monitors incidents, assigns root cause analysis tasks and sends defect resolution notification back to the test manager. The test coordinator is the focal point for communication with the test manager.
- ◆ **Senior business manager**—Responsible for reviewing the monthly status report with the project manager. The senior business manager usually reports to the project sponsor.
- ◆ **Senior delivery manager**—The person who owns a delivery budget and whose signature is required on the statement of work.
- ◆ **Project sponsor**—Major stakeholder who is responsible for the business success of the project, specifically, ensuring that the business objectives for which the project has been undertaken are met. The project sponsor is the owner of the overall project.

Inputs/Outputs

The Closing Frame processes interact with the rest of the project frames via entry and exit points, as shown in the Closing Frame process flow diagram in Figure 16.1:

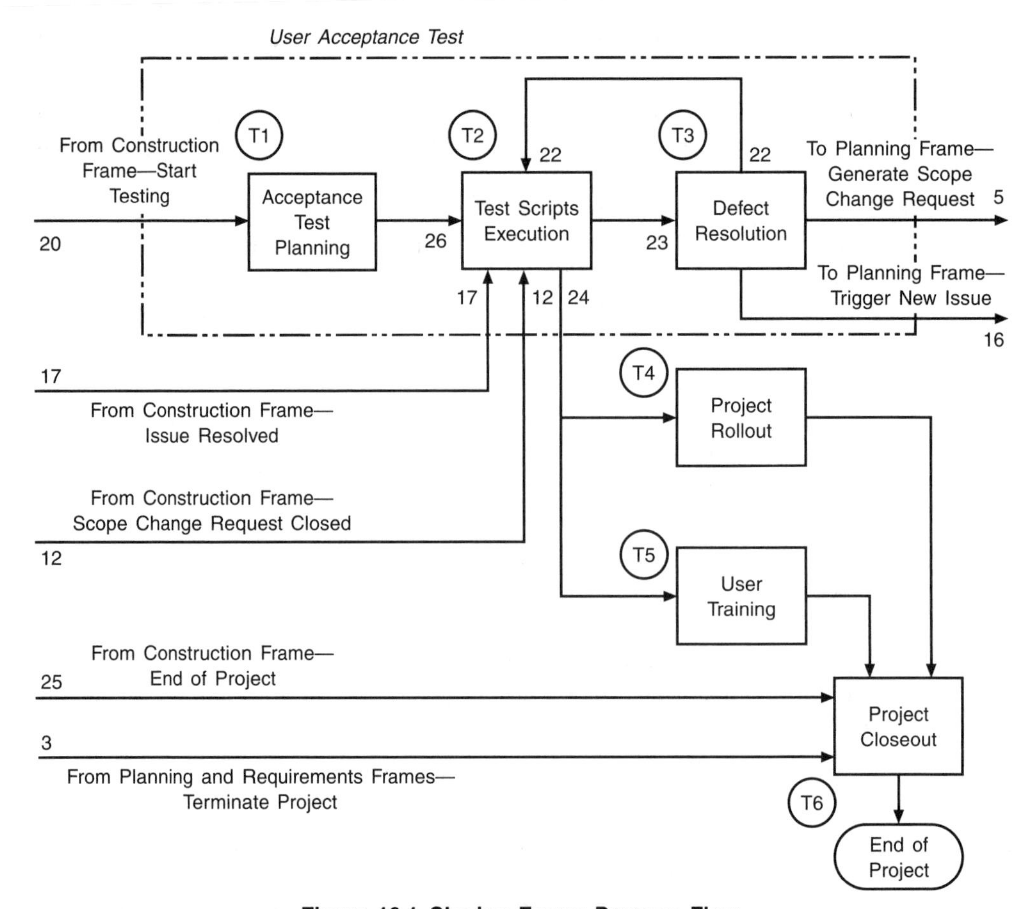

Figure 16.1 Closing Frame Process Flow

- **Exit point 15**—To the Planning Frame to trigger a scope change request
- **Exit point 16**—To the Planning Frame to trigger a new issue

- **Entry point 3**—From the Planning and Requirements frames to terminate the project
- **Entry point 12**—From the Construction Frame with notification that the scope change request is complete
- **Entry point 17**—From the Construction Frame with notification that the issue has been resolved
- **Entry point 20**—From the Construction Frame to start acceptance testing
- **Entry point 25**—From the Construction Frame to the end of the project

Closing Frame High-Level Process Flow

The Closing/Testing Frame consists of the high-level processes shown in Figure 16.1, which are decomposed into the following detailed processes:

1. **Acceptance Test Planning (T1)**—The purpose of this process is to detail the steps necessary to run the acceptance test. The process includes test planning, test scripts development and defect management planning. This is a business-owned process.
2. **Test Scripts Execution (T2)**—This process tests functionality of the product developed in accordance with the scenarios or scripts developed. This is a business-owned process.
3. **Defect Resolution (T3)**—The purpose of this process is to identify the steps necessary to fix a defect by performing root cause analysis and the steps to track defects that occur before and after release of the product into production. This is a delivery-team-owned process.
4. **Project Rollout (T4)**—The purpose of this process is to establish methods for project rollout and to roll out the project to the production environment or deliver the product to users.
5. **User Training (T5)**—The purpose of this process is to identify training needs and establish user training.
6. **Project Closeout (T6)**—The purpose of this process is to close project documentation, release resources and analyze the finished project in order to identify successes and failures during project execution so as to be able to repeat successes and avoid failures in the future.

The process is initiated when the request to start testing comes from the Construction Frame via entry point 20. The Acceptance Test Planning (T1) process is executed first, where scripts are developed and required tests are scheduled. From there, the process flow moves to entry point 26 for the Test Scripts Execution (T2) process, where the next scheduled script execution tests a specific element of the product functionality to confirm compliance with the corresponding business requirement.

If a deviation from the expected results, also called an incident, occurred during the script execution, the process flow enters the Defect Resolution (T3) process via entry point 23. When defect resolution is completed, the process flow will loop back to the Test Scripts Execution (T2) process to start execution of the same test script again to verify that the defect was resolved. In fact, as can be seen in the detailed process flow of the Defect Resolution (T3) process (see Figure 16.5), the process flow may loop back to process T2 even before the end of the defect resolution process.

If the scheduled script execution is completed without incident, the next scheduled script will be executed, until there are no more scripts left to run, in which case the user acceptance test is complete and the process flow via exit point 24 enters the Project Rollout (T4) process and the User Training (T5) process at the same time. During execution of process T4, deliverables are handed over to users in the production environment or warehouse. Process T5 addresses developing training materials and conducting user training. When both processes are complete, the process flow enters the Project Closeout (T6) process, which must receive end-of-project notification via entry point 25 in order to be executed. During execution of process T5, sign-offs are received, resources are released and all project documentation is archived. The project will be formally closed and the entire project process flow analyzed. Conclusions will be drawn about mistakes that occurred and how they could have been avoided. Upon closing the project, the project flow ends.

Process T6 will also be executed, even without user acceptance testing, if a terminate project notification comes from the Requirements Frame or the Planning Frame via entry point 3.

High-level processes T1, T2 and T3 represent the set of user acceptance test processes. In the following sections, those processes are broken down into detailed sets of processes.

User Acceptance Test Overview

The user acceptance test is a set of user-developed and -conducted tests to confirm that a completed product satisfies business requirements as outlined in the business requirements document. Each individual test is conducted according to the script of test execution developed, which is the detailed step-by-step description of test activities. Each script represents a plan to test a specific business feature of the product in an environment that is as close as possible to real-life operating conditions. The expected result is clearly described in each script, and the test is rated pass or fail. All scripts must be documented in the script and incident repository.

The user acceptance test scripts are developed and the tests conducted by the acceptance test team, which consists of business users assigned by the business. The acceptance test support team, assigned by the delivery organization, will assist business users in that task. Their responsibility is also to resolve defects.

The user acceptance test procedure described here may not be applicable to all projects, such as construction, process improvement, etc. For example, construction projects usually have their own test methods which are compliant to construction, electrical and other codes, as well as their own approval chain. It is not possible to provide specific user acceptance test methods applicable to all project types. Therefore, the processes described here provide general user acceptance test steps and guidelines, but no specific examples of implementation. Some process names, such as Identify Test Conditions for Each Requirement in Business Requirements Document or Identify Test Cases for Each Requirement in Business Requirements Document, give the general idea of this concept, but no examples or details are provided, since details are very specific for every business or practice area.

The following will be achieved during the user acceptance test:

- There is a clear understanding of the testing scope.
- The comprehensive test plan is developed, which provides the schedule, reporting requirements, analysis and tracking activities.
- Test scripts are developed to provide steps and conditions for testing of product functionality, one test script for one element of functionality.
- All business-related functions and requirements, as described in the business requirements document, are tested.
- Product performance is tested.
- Product interfaces with other products and users are tested.

- The product tests use actual work scenarios.
- All product defects found during script execution which affect product functionality are fixed and documented.
- A root cause analysis of every defect is performed and documented, unless agreed otherwise by all parties.
- Nonfunctional requirements, such as safety, security and others, are met.
- Every possible effort is made to reduce the number of defects discovered after release into production.
- Training plans are developed based on actual user operations.
- A communication plan is developed and implemented for communication among the delivery team, client, business users and management during all test activities.
- All change requests are identified and prioritized.

The most complicated and comprehensive test scripts are developed for projects in the software, information technology and to a lesser degree electronics and electromechanical fields. Those scenarios incorporate various types of inputs, including intentionally erroneous, while testing expected outputs. They often simulate unexpected inputs and try to break the existing project logic. Therefore, many details of test scripts described below are applicable only to specific business areas.

The user acceptance test includes the following high-level sequential processes:

1. Acceptance Test Planning (T1)
2. Test Scripts Execution (T2)
3. Defect Resolution (T3)

Defects

A defect is a quality gap between a specified quality characteristic and the actual quality, which results in a product or service not satisfying its intended use.

Defects are identified when script execution conditions are not met successfully. The most common reasons for defects discovered during acceptance testing are:

1. A business requirement not appropriately defined or documented, resulting in initiating a change request
2. A business requirement not correctly implemented due to misunderstanding or errors in design or implementation
3. Requirements cannot be fully implemented with the technology or materials selected
4. A misunderstanding or mistake in testing
5. Incorrect setup of the test environment
6. Insufficient communication between the delivery team and business
7. Lack of the required skills in the delivery team
8. Lack of the required skills in the acceptance test or acceptance test support team
9. Project planning errors

10. Not all quality audits and reviews performed during the course of the project
11. Incorrect assumptions made during design or testing

A defect must be corrected before continuing script execution. A defect is a problem that will consistently show up in a similar situation in every script execution. For example, a bug in a software program code or a design error in a mechanical/electrical assembly is a defect.

Sometimes an incident may be defined as a defect only after multiple occurrences. If, for example, the tester complains about a perceived problem, which turns out to be an incorrect use of a test tool, this is not a defect. However, if more than one tester complains about the same problem, then there may be a true defect in the test tool training. In order to reduce the number of future defects, every defect requires a root cause analysis. The assigned delivery team members will perform a root cause analysis when an incident is determined to be a defect. The purpose of root cause analysis is to figure out what is really causing the defect to occur and remove it, so the situation does not occur again, rather than simply continuing to deal with the symptoms. If an incident is not a defect, then it is an issue, which must be resolved in the framework of issue resolution.

Every occurrence of a problem must be documented in the script and incident repository, along with the root cause analysis.

Quality

The testing process does not build quality into a product. Testing is a means of determining the quality of the product tested and removing the defects found. Since it is often impractical to test every single detail due to schedule and cost constraints, some defects may be discovered after the system is released into production, if quality is not built in.

According to the Project Management Institute's definition, quality is the characteristic of an entity that affects its ability to satisfy stated or implied business needs and its fitness for use. The quality management process embeds quality into the project and thus the product from day one of project development. By the time testing of the product is performed, it is way too late to improve the quality of the deliverable.

Defects may be found any time during the project life cycle and not only during the user acceptance test. The cost of defect repairs increases proportionately the later in the project life cycle a defect is detected. Correcting a problem in the Requirements Frame may not cost much. Correcting a problem postimplementation may cost tenfold and more.

V-Model

The V-model, shown in Figure 16.2, is the most commonly used approach to testing. The V-model represents the project life cycle. It shows the various stages in development and testing, as well as relationships between various stages. The verification or validation of test activities relates to their corresponding requirements or specifications. The related

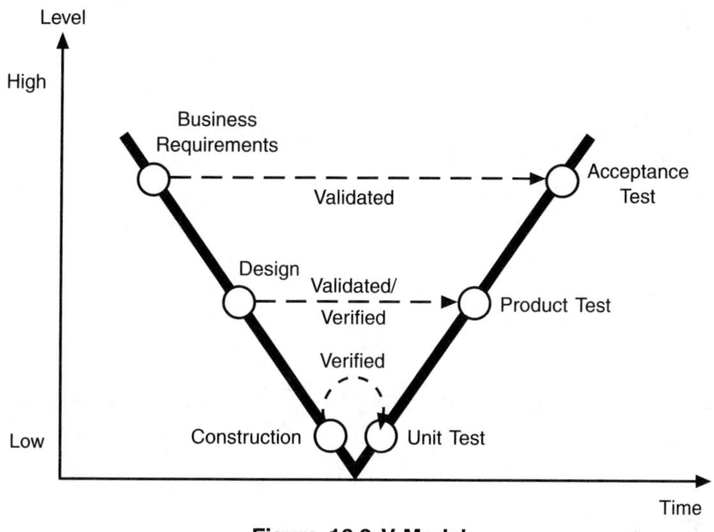

Figure 16.2 V-Model

testing activities to ensure that the right functions are performed are called *validation* activities; activities to ensure the correct or reliable performance of those specified functions are called *verification* activities. If validation of activities is performed on a specific level of detail, the verification of activities will be performed on exactly the same level of detail and vice versa.

Every product under development consists of a number of elements, called units. Thus, in a software project, a product may be called a system and software modules called units. In a mechanical engineering project, a product may be called an assembly and units are subassemblies.

At the top level of the project view, the business requirements established at the beginning of the project may be validated by the acceptance test, which is developed and performed by the business at the end of the project. At a lower or more detailed project level, the product design produced at the initial stages of the project first must be validated and then verified by the product test. The product test is developed by an analyst or engineer and performed by several members of the delivery team at the last stages of the project before the product is scheduled for handover to the client for acceptance testing.

At the lowest and most detailed level, during project construction, unit verification is done by the unit test, which is developed and performed immediately by the person who created the unit. A unit is the smallest testable part of the product. A unit test is performed by the individual developer who developed that unit/module/subassembly. It verifies that the unit behaves exactly as required when tested independently of other units.

The next higher level of testing is the product test, when all tested units are assembled together. For a software development project, the product test combines the system and integration tests. The product test will verify and then validate the existing quality elements:

- Correctness
- Reliability
- Usability
- Maintainability
- Testability
- Reusability

Both unit and product tests are very specific for each business area and product type. In order to understand the unit and product testing methods and processes, one must be a technical specialist in that specific area. Therefore, the above tests are out of the scope of this book.

Remember that the user acceptance test is a business requirements validation process conducted by business users.

Script and Incident Repository

All scripts and every incident must be documented. There are many off-the-shelf tools which do this. However, if for any reason such a tool is not available, it is possible to build the repository in the same way the project control book is built, which is an example of a document repository. Expanding the project control book in order to include scripts and incidents is not recommended, because the project control book is a tool used mostly by the project manager and the delivery team, whereas the script and incident repository is mainly a user tool for user acceptance testing. Script and incident entry forms are provided later in this chapter.

Acceptance Test Planning (T1)

Test activities must be planned, like any other project activities, so that the test execution can be estimated, scheduled and tracked. Keep in mind that some test activities, which must be planned, are carried out by members of the client team, but they are not chargeable to the client and are not a part of the budget. However, if those activities are not done in accordance with the plan, then the delivery team activities will cost extra when supporting the client team test tasks. If this happens, a scope change request must be submitted, as discussed in Chapter 10.

In order to plan test activities, a good understanding of those activities, a detailed schedule/calendar, the script sequence and dependencies must be entered in the overall project plan. The planning process must allocate the following planning tasks for implementation in the overall project plan:

- Approach to testing
- Test methods
- Risk assessment
- Detailed list of acceptance test tasks
- Task dependencies

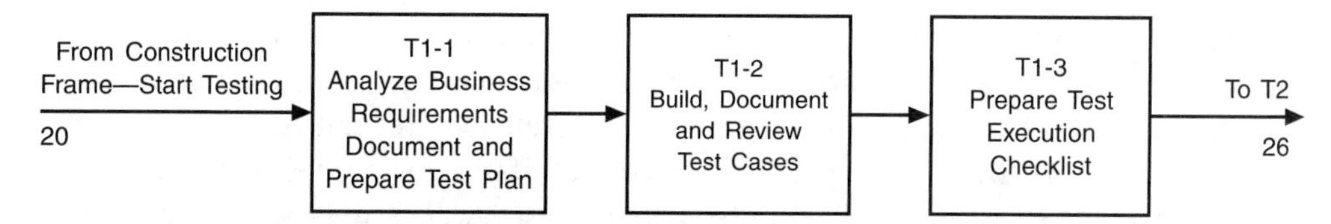

Figure 16.3 Acceptance Test Planning (T1) Process Flow

- Resource assignment
- Acceptance test schedule/calendar
- Communication plan during the acceptance test
- Acceptance test exit/completion criteria
- Test cases and test scripts

The Acceptance Test Planning (T1) process, shown in Figure 16.3, consists of the following subprocesses:

1. Analyze Business Requirements Document and Prepare Test Plan (T1-1)
2. Build, Document and Review Test Cases (T1-2)
3. Prepare Test Execution Checklist (T1-3)

Analyze Business Requirements Document and Prepare Test Plan (T1-1)

This process is decomposed into the following elementary processes:

1. Analyze Business Requirements Document (T1-1-1)
2. Identify Test Conditions for Each Requirement in Business Requirements Document (T1-1-2)
3. Identify Test Cases for Each Requirement in Business Requirements Document (T1-1-3)
4. Break Down Each Test Case for Detailed Elementary Test Steps (T1-1-4)
5. Document Test Cases (T1-1-5)

Build, Document and Review Test Cases (T1-2)

The acceptance test cases are developed to validate that the system is performing in accordance with the business requirements document. Test cases are comprised of inputs, transactions and outputs.

Inputs. Requirements to be tested based on the business requirements document should be identified, and a combination of input scenarios to test all requirements should be documented. Attempts should be made to "break" the system by setting up illogical conditions and invalid data entry, such as entering smaller than minimum values, larger than maximum values, nonnumeric values where numeric values are expected or nonalpha characters in the customer name field. The expected result from the incorrect data entry is rejection by the system, but the system should never crash.

Transactions. Each business requirement must be tested as specified in the business requirements document by generating several transactions with both valid and invalid conditions. The transactions should cover the complete requirement and then test what happens if the order of their execution is incorrect. If a transaction deletes some data, the data should be restored upon completion of the test case, because test cases developed by other testers are based on a predetermined state of the data. Under no circumstances should the system crash; rather, it must provide an acceptable and informative error message.

Outputs. Expected results should identify the following elements:

- Results of each step in the test script—If expected results are numeric values, the range must be precisely identified or calculated.
- System status during script execution and upon its completion, such as the ability or inability to run some transactions, display screens, etc.
- Data modifications during a test script execution—As mentioned earlier, after a script execution is complete, data must be restored to the state before the script execution, unless otherwise planned.

Prepare Test Execution Checklist (T1-3)

Before running tests, the following checklist must be used to verify readiness:

- The test case execution schedule is available
- The test environment is ready
- Resources are committed
- Backup and recovery procedures for software projects are in place
- The test data, if any, is loaded and accessible

Regression testing is retesting of previously working test scripts following defect resolution to ensure that new defects have not been introduced as a result of the changes made. Typically, regression testing is required if modifications have been introduced to an existing product or to the environment. In this case, tests should be run to ensure that all existing processes that may possibly be affected still work after the change. If the original script was executed and closed, a copy of the script should be run. In this case, the script number will consist of the original script number, followed by a hyphen and then the copy number. For example, if the original script number was 100, then the first copy will be 100-1, the second copy 100-2, etc.

Before beginning test script execution after defect removal, it must be determined whether a regression test is required. If it is, the regression test must be conducted after removing defects before the subsequent script execution.

Test Scripts Execution (T2)

The Test Scripts Execution (T2) process is decomposed into the processes listed in Table 16.1. The process flow is shown in Figure 16.4. Prior to executing process T2, all scripts

Table 16.1 Test Scripts Execution (T2) Process Decomposition

Process #	Process Name	Owner/Executor
T2-1	Open Next Script	Test team member
T2-2	Execute Script	Test team member
T2-3	Approve Incident	Test team member
T2-4	Open New Incident Report	Test team member
T2-5	Document Script Execution	Test manager
T2-6	Receive Defect Resolution Notification	Test manager
T2-7	Close Incident	Test manager
T2-8	Reschedule Test	Test manager

must be entered in the script and incident repository using the script entry form provided in Table 16.2. The process starts when the Open Next Script (T2-1) process is triggered by one of the following:

1. The test execution schedule, indicating which specific script must be executed next.
2. The Reschedule Test (T2-8) process, which means that the defect that occurred during the previously executed script has been fixed. In this case, the schedule also must indicate the correct timing to rerun the script and run regression tests.
3. After the previous script is successfully completed and documented in the Document Script Execution (T2-5) process and the control point question (acceptance test complete?) is answered no, the process flow loops back to the Open Next Script (T2-1) process, starting execution of the next script in the queue.
4. After an incident has been approved by the test manager in the Approve Incident (T2-3) process and while a new incident is being opened in the Open New Incident Report (T2-4) process, the next script in the queue is opened for execution. Incident approval by the test manager is required in order to reduce the number of false defects, when it is clear that perceived defects are not defects but rather problems not directly related to the quality of the deliverable.

The process retrieves the next scheduled script information from the script and incident repository, and the script is executed step by step in the Execute Script (T2-2) process until execution is complete or an incident occurs during execution. If the script execution is successful, the answer to the control point question (execution successful?) is yes, and the process flow is directed to the Document Script Execution (T2-5) process, where the test manager is notified, the script is closed and the script information is stored in the script and incident repository. At the next control point question (acceptance test complete?), the process flow loops back to the Open Next Script (T2-1) process if the answer is no, which means that the user acceptance test of the product is not yet complete and the next scheduled script must be executed. Otherwise, if the answer is yes, the process flows to the Project Rollout (T4) process and the User Training (T5) process, which are outside of the script execution process.

If the answer to the control point question (execution successful?) is no, then the test manager approves opening the incident in the Approve Incident (T2-3) process, and

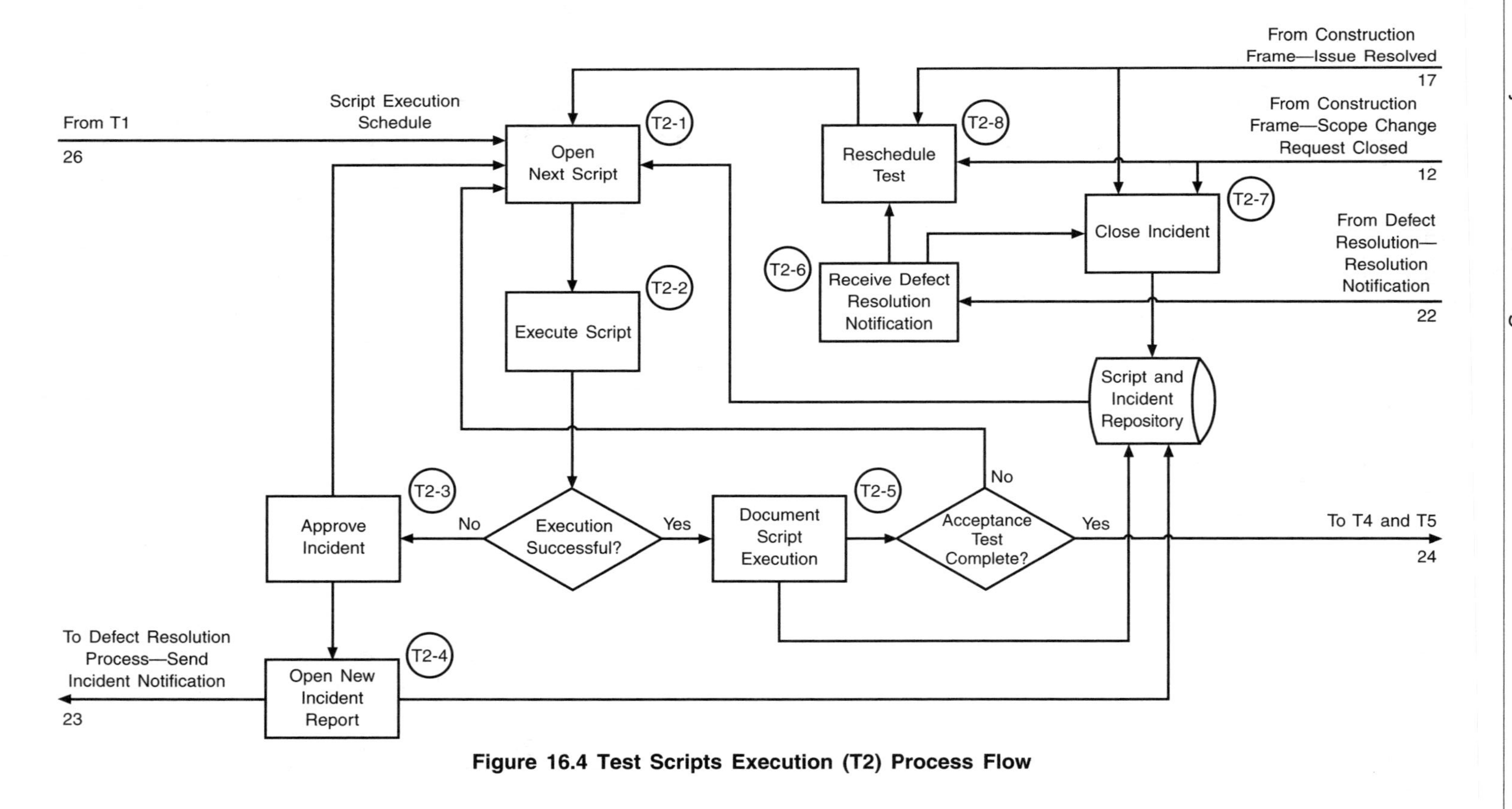

Figure 16.4 Test Scripts Execution (T2) Process Flow

Table 16.2 Script Entry Form

Script Entry Form	
Project Name:	
Project Manager:	
Test Manager:	
Test Coordinator:	
Script Author:	
Script Owner:	
Script Number:	Execution #:
Script Status: ☐ Pending ☐ Defect Resolution ☐ Completed ☐ Escalated	
Script Title:	
Script Description:	
Script Open Date:	Script Close Date:
Closing Status: ☐ Ended with No Incidents ☐ Defect Fixed ☐ New Scope Change Request Opened ☐ New Issue Opened	
Products/Projects Impacted: ☐ None 1. ______ 2. ______ 3. ______	
Script Comments:	
Step #1 Assigned to: ______ Step Status: ______ Description:	
Step Expected Results:	
Step Actual Results:	
Step Comments:	
Step #2 Assigned to: ______ Step Status: ______ Description:	
Step Expected Results:	
Step Actual Results:	
Step Comments:	
Continue with Steps 3, 4, 5, etc.	

a new incident report is opened in the Open New Incident Report (T2-4) process. The incident entry form in Table 16.3 is filled out and stored in the script and incident repository. The incident number consists of the script number, a hyphen and the step number. At the same time, an incident notification is sent to the test coordinator via exit point 23 to assign a test support team member to resolve the incident and perform root cause analysis, which is done in the Defect Resolution (T3) process. From there, the flow loops back to the Open Next Script (T2-1) process to start execution of the next script.

Table 16.3 Incident Entry Form

Incident Entry Form	
Project Name:	
Project Manager:	
Test Manager:	
Test Coordinator:	
Incident Owner:	
Incident Number:	
Script Number:	Step Number:
Incident Open Date:	Incident Close Date:
Resolution Priority: ☐ Immediate ☐ 1 to 2 Weeks ☐ Delayed Pending Notification	
Script Title:	
Status: ☐ Owner Assigned ☐ Opened ☐ Closed	
Incident Description:	
List of Incident Attachments:	
Reasons for Closing: ☐ Defect Removed ☐ New Scope Change Request Opened ☐ New Issue Opened Incident Closing Comment:	
Defect Root Cause Analysis	
Defect Description:	
Analysis Description:	
Performed by:	Date:

An incident will only be approved by the test manager when he or she is certain that the incident did not occur due to a tester error, the test data is correct, there was no power failure, etc. Otherwise, the problem must be fixed first. Then the process flow will loop back to the Open Next Script (T2-1) process to continue the same script execution when possible or start the script from the beginning. If the incident is approved, process T2-4 opens a new incident report and then follows the flow as described above.

When the defect is resolved, the resolution notification comes from the Defect Resolution (T3) process via entry point 22. The process flow enters the Receive Defect Resolution Notification (T2-6) process, where the test manager receives notice and confirms with the defect owner that the defect was indeed resolved. The incident will be closed in the Close Incident (T2-7) process and documented in the script and incident repository. At the same time, the script execution will be rescheduled in the Reschedule Test (T2-8) process, and a decision will be made whether regression tests are required.

During the Defect Resolution (T3) process, a new scope change request or a new issue may be opened. When the scope change is completed or the issue is resolved, the corresponding notification comes via entry point 12 or entry point 17, respectively, to the Close Incident (T2-7) process and also to the Reschedule Test (T2-8) process in order to rerun the failed script again.

The script owner is responsible for the initial data entry in the incident entry form (Table 16.3). It is also the script owner's responsibility to keep the script data in the script entry form (Table 16.2) current, updating it daily if new information is available.

Script Entry Form

The following information must be provided in the script entry form (Table 16.2):

1. **Project name**
2. **Project manager**
3. **Test manager**
4. **Test coordinator**
5. **Script author**—The person who developed the script.
6. **Script owner**—The acceptance test team member who is assigned responsibility to execute the script.
7. **Script number**—An example of a script number is 215-02. The first part of the script number is the sequential number assigned to the script. The second part is the execution number; for the first run of a script, it is 01 and increases for each subsequent run. Under normal circumstances, a script is run only once. If an incident occurs during a script run, after taking care of the problem, the script is executed again.
8. **Execution #**—Indicates whether the script is being run the first time, the second time, etc.
9. **Script status**—The status of the last step executed. Upon entering script information, the status is pending (script is being executed), but later may be defect resolution (incident occurred, which is being resolved), completed (script execution is successfully completed) or escalated (the incident has been escalated).
10. **Script title**—Should relate to the business function being validated.

11. **Script description**—A detailed description of the script.
12. **Script open date**—The date when the script was opened.
13. **Script close date**—The date when the script was closed.
14. **Closing status**—Status at the end of execution may be ended with no incidents, defect fixed, new scope change request opened or new issue opened.
15. **Products/projects impacted**—Products or projects that may be impacted by this script execution.
16. **Script comments**—Any helpful information to understand the script.
17. **Step #**—Each script may have a number of steps, which are sequentially numbered, beginning with 1.
18. **Assigned to**—The acceptance test team member who is assigned responsibility to execute the step. Usually, the step owner is the same as the script owner.
19. **Step status**—See script status.
20. **Step description**—A description of the step executed, unless already included in the script description.
21. **Step expected results**—A description of the expected results, which must include expected resulting numbers based on the input and numeric processing, names of files with expected screens, etc.
22. **Step actual results**—Actual results of the script execution. This information is entered at the time of script execution. If variance between expected and actual results is acceptable, the script execution rating is pass. Otherwise, it is fail.
23. **Step comments**—Any useful information to understand the step.

Incident Entry Form

The entries in the incident entry form in Table 16.3 are as follows:

1. **Project name**
2. **Project manager**
3. **Test manager**
4. **Test coordinator**
5. **Incident owner**—The test support team member in charge of resolving the incident.
6. **Incident number**—The incident number consists of a sequential number, which always begins with 1 for each new script.
7. **Script number**—The sequential number assigned to a script at the time of script generation.
8. **Step number**—Each script may have a number of steps, which are sequentially numbered, beginning with 1.
9. **Incident open date**—Date when the incident was opened.
10. **Incident close date**—Date when the incident was closed. An incident will be closed when one of the following happens: the defect was resolved, a new scope change request is opened or a new issue is opened.
11. **Resolution priority**—Priority is assigned at the time the incident is opened. There are three priority levels: immediate, 1 to 2 weeks and delayed pending notification.

12. **Script title**—Should relate to the business function being validated.
13. **Status**—The incident status may be one of three choices: defect owner assigned (defect being resolved), opened or closed.
14. **Incident description**—Detailed description of the problem.
15. **List of incident attachments**—A list of attachments needed to understand the problem.
16. **Reasons for closing**—An incident may be closed due to one of the following reasons: defect removed, new scope change request opened or new issue opened.
17. **Incident closing comment**—Comments required to understand the reason for closing. If the reason for closure is a new scope change request or issue opened, it must be explained here.
18. **Defect root cause analysis**
    - ◇ **Defect description**—If different from the incident description above, the information should be entered here.
    - ◇ **Analysis description**—The root cause analysis report.
    - ◇ **Performed by**—The test support team member who performed the root cause analysis.
    - ◇ **Date**—Date of the root cause analysis.

Defect Resolution (T3)

The Defect Resolution (T3) high-level process is decomposed into the subprocesses listed in Table 16.4. The process flow is shown in Figure 16.5.

This process starts when an incident notification comes from the Test Scripts Execution (T2) process via entry point 23. The test coordinator receives an e-mail notification in the Get Incident Notification (T3-1) process. The incident information is pulled from the script and incident repository and reviewed with members of the test support team in the Review Incident (T3-2) process, where a detailed analysis is done to establish whether the incident is a defect. If the incident information is insufficient for further action, the answer to the control point question (information complete?) is no, and a request is made to the test manager to provide additional information in the Clarify Information (T3-3) process. When clarification is received, the process flow loops back to the Review Incident (T3-2) process and the incident is reviewed again. If the answer

Table 16.4 Defect Resolution (T3) Process Decomposition

Process #	Process Name	Owner/Executor
T3-1	Get Incident Notification	Test coordinator
T3-2	Review Incident	Test coordinator
T3-3	Clarify Information	Test coordinator
T3-4	Assign Defect Owner	Test coordinator
T3-5	Resolve Defect	Test support team member
T3-6	Track Resolution	Test coordinator
T3-7	Escalate	Test coordinator
T3-8	Perform Root Cause Analysis	Test support team member
T3-9	Notify Test Manager	Test coordinator

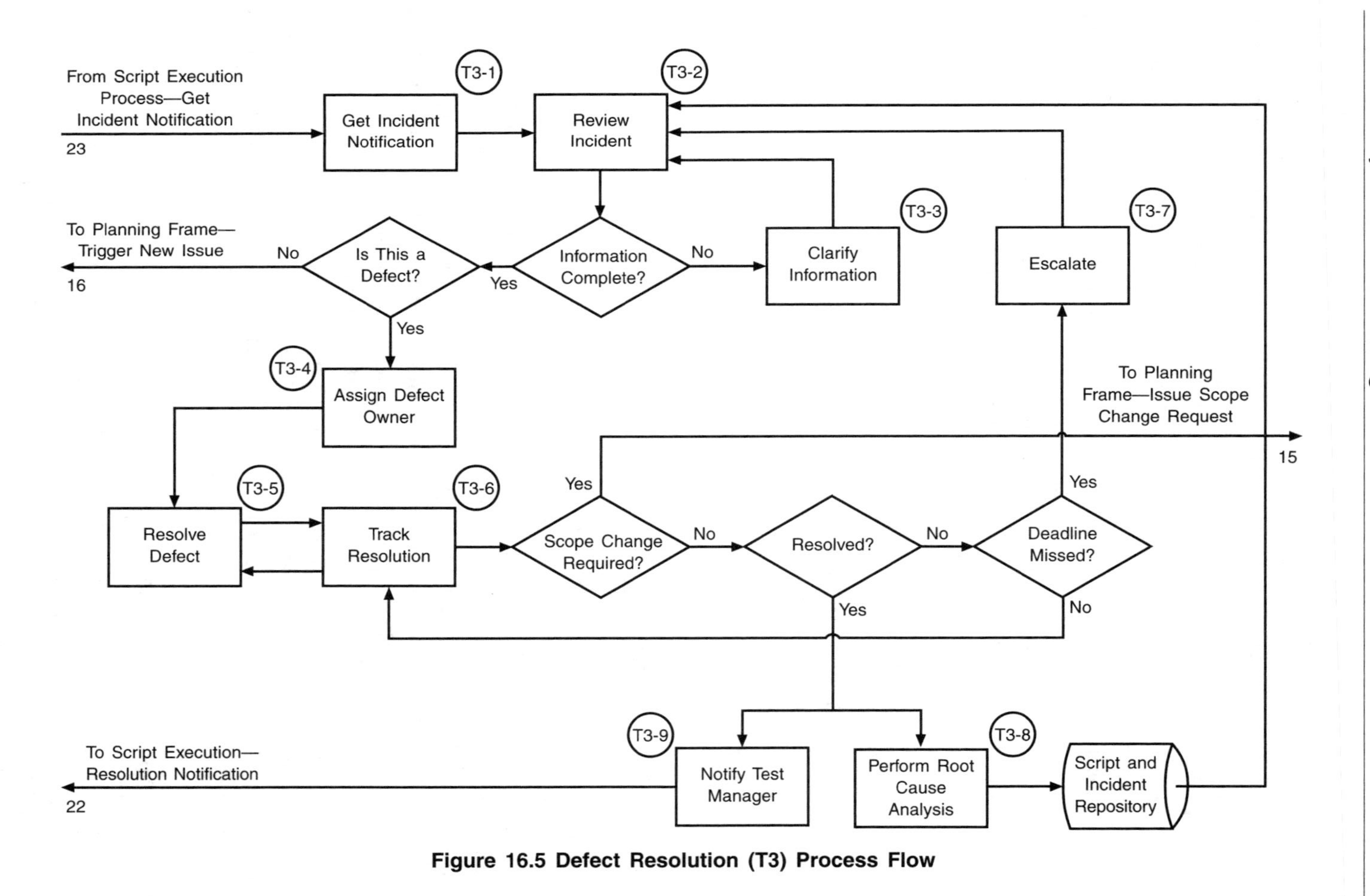

Figure 16.5 Defect Resolution (T3) Process Flow

is yes, the next control point question (is this a defect?) is reached. If it is determined that the incident is not a defect, then it is an issue that must be resolved using the issue resolution process described in Chapter 8. In this case, the answer is no, and the request is forwarded to the Planning Frame via exit point 16 to open a new issue. If the answer is yes, then the incident is a defect and the process flow enters the Assign Defect Owner (T3-4) process, where the test coordinator assigns a member of the test support team to fix the defect. While the defect is being fixed in the Resolve Defect (T3-5) process, progress will be tracked by the test coordinator in the Track Resolution (T3-6) process. If the defect cannot be resolved without a scope change to the project, then the answer to the control point question (scope change required?) is yes, and a request is issued to the Planning Frame to issue a new scope change request via exit point 15. Otherwise, the next control point question (resolved?) is reached. If the defect has not been resolved yet, the answer is no, and the process flow reaches the next control point question (deadline missed?). If the deadline for defect resolution has not been missed, the answer is no, and defect resolution tracking continues by looping back to the Track Resolution (T3-6) process. However, if the deadline has been missed, the answer is yes, and the problem is escalated to the project manager in the Escalate (T3-7) process. If the project manager is not successful in remediation of the problem, he or she must escalate it to the senior delivery manager.

After escalation, the process flow loops back to the Review Incident (T3-2) process, where the test coordinator reviews the incident again and the defect resolution starts all over again, as described above. Going back to the Review Incident (T3-2) process, rather than continuing execution of the Resolve Defect (T3-5) process, provides the opportunity to assign a different defect owner in process T3-4.

NOTE It is always a good idea to assign a different test support team member for resolution of a specific incident, even if the currently assigned team member did not make any mistakes. This is done to alleviate tension within the business team due to delay and to demonstrate that the delivery team is making a visible effort to resolve the problem.

When the defect is fixed, the answer to the control point question (resolved?) is yes, and the flow enters the Notify Test Manager (T3-9) process, where the test schedule is updated to allow the failed script to run again. Process T3-9 also sends a resolution notification to the Test Scripts Execution (T2) process via exit point 22, so that the failed script may be run again in accordance with the new script execution schedule. At the same time, when the test manager is notified that the defect is fixed, the process flow enters the Perform Root Cause Analysis (T3-8) process, where the root cause of the defect is determined. This process is required to avoid future problems due to the same reason and to build a knowledge database, which may assist in quick resolution of similar problems in the future. The root cause analysis may take up to 72 hours of investigation. At the end of process T3-8, the root cause analysis is documented in the script and incident repository.

Metrics

During script development and execution, the following metrics, at a minimum, will be recorded for eventual storage in the project control book:

- Number of test scripts and test steps planned and prepared
- Number of test scripts and test steps planned for execution to date
- Number of test scripts and test steps executed to date
- Number of defects discovered and fixed for each test script
- Total number of defects discovered and fixed

In addition, the effort and duration of the test planning, preparation and execution stages, as well as defect resolution and root cause analysis, must be recorded and tracked.

Project Rollout (T4)

The objectives of this process are to ensure that:

1. The user acceptance test has been successfully completed and the deliverable product meets the documented business requirements.
2. The product is moved/installed and is fully functional in the user or production environment.
3. User acceptance/approval is received and documented.
4. The client receives all required authorizations to use the product on a daily basis or to transfer or sell the product to a third party.

Upon completion of the user acceptance test, evidence must exist that all scripts run successfully, all incidents are closed and no outstanding defects, scope change requests or issues exist. The test manager must ensure that all scripts are successfully executed and documented in the script and incident repository. This provides evidence that the user acceptance test has been successfully completed.

Moving the product to the user environment involves methods specific to each industry. For example, in the information systems area, the product moves to the production library, with new sets of programs, libraries and databases initialized with the production data. The cutoff to the production environment takes place on a planned date and time. In addition, security arrangements must be made to ensure access to the product by end users, at the same time relinquishing the delivery team's access, except for maintenance personnel. Other business areas use other methods to ensure that the product will meet business requirements when operated in the user environment by the assigned end users of the product.

It is at this point that the product of the project moves to the steady state. For sustainable design, it is important that the project manager keep this moment in mind throughout the Planning and Construction frames. By doing so, the project manager is more aware of long-term effects. For example, if the product of the project is a single-serve coffeemaker, the recyclability of the coffee packets used in everyday operation is thought through at an early enough point to have an impact. It is important for a project manager to think this way to align the sustainability and environmental attitudes of the corporation with the project itself. This may or may not be applicable to software products, but it is always good to keep the sustainability aspect in mind.

Just before transferring authority to the client, the project manager will forward a project acceptance certificate (Table 16.5) to the client for approval. Once approved, the certificate will serve as a basis for administrative and financial closure of the project. The project acceptance certificate, along with the approval documentation, will be stored in the project control book. From that moment on, the delivery organization will provide warranty repairs in accordance with the documented agreement with the client.

If the client withholds project approval for any reason, an attempt is made by the project manager to resolve the issue with the client. An action plan is developed to resolve the issue by a certain deadline. If approval is still not received by the deadline,

Table 16.5 Project Acceptance Certificate Form

Project Acceptance Certificate	
Delivery Organization:	
Client Organization:	
Project Name:	
Client Name:	
Project Manager:	
Submitted Date:	Rejected Date:
Reason for Rejection:	
Resubmitted Date:	Rejected Date:
Reason for Rejection:	
Accepted Date:	Accepted by:
Deliverables List:	
Comments:	

the project manager will escalate the issue to the senior delivery manager, who in turn will escalate it to the senior business manager and from there to a higher level of management, arbitration or a court of law. As long as there is no approval, the client and business users will have no authority to use or operate the product.

User Training (T5)

When end users are well trained in the use of the product, there are two obvious benefits:

- The delivery organization will save time and expenses by limiting false warranty service calls.
- The client organization will increase productivity and reduce product downtime.

Therefore, it is important to plan and conduct the end-user training in a way that establishes the fastest possible learning curve and users gain hands-on practical experience.

The User Training (T5) process, as shown in Figure 16.6, can be decomposed into the following subprocesses:

1. Obtain List of Trainees (T5-1)
2. Assess User Needs (T5-2)
3. Develop Training Materials (T5-3)
4. Deliver Training (T5-4)

Obtain List of Trainees (T5-1)

A list of trainees may be obtained from managers of those business units that are planned to operate the product developed. Those business units or their representatives must

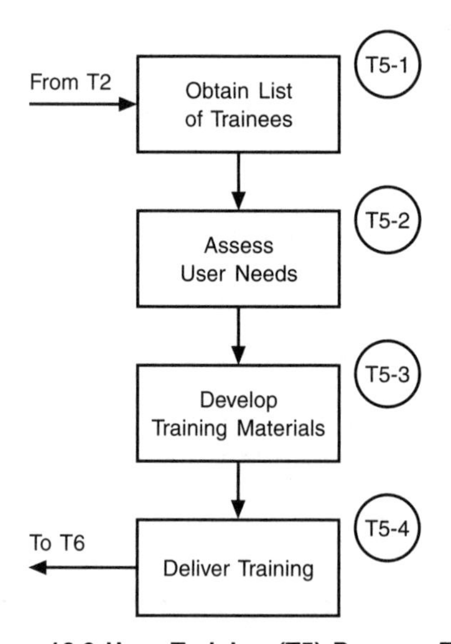

Figure 16.6 User Training (T5) Process Flow

have participated in developing the business requirements and, preferably, performed user acceptance tests.

When the list of trainees is received, it should indicate how many users can participate in training at the same time without affecting daily operation of the business units. This information must be taken into consideration when identifying training methods and scheduling training.

If the delivered product is a brand new product, then end users will have to learn everything from the business background to practical use of the product until they attain satisfactory proficiency in using it. Sometimes a newly developed product replaces an existing product, which upon successful implementation will increase performance, accuracy, etc. This situation may require less training, but not always. If the end users developed specific habits in operating the old product and the new method of operation is substantially different, the upgrade may be difficult to learn.

Assess User Needs (T5-2)

The assessment of user needs may require study of incidents that occurred during the user acceptance test due to user errors and evaluation of their skills. The assessment may require producing checklists to understand end-user readiness to learn both the business part as well as the product operation.

Always keep in mind that users have different skills, different education and different attitudes. While it is not always practically possible, it may be beneficial to provide at least two levels of training to avoid novice users getting confused or advanced users becoming bored.

Sometimes different groups of users use different features of the product. In that case, separate training is mandatory for each group of end users.

Develop Training Materials (T5-3)

Training materials will depend on the method of training delivery and the evaluation of user needs, as well as the number of trainees. Development of training materials is best achieved by the combined effort of the delivery team and the business team members, so that there is both technical and business input into the training. Therefore, before beginning this task, a team for development of training materials must be formed that reports to both the project manager and the client.

The most effective method of training is an individual hands-on instructor training a group of 15 to 20 users. Unfortunately, this is the most expensive. The least effective and the least expensive method of training is providing users with the product manual and hoping they figure it out by themselves. Other training methods include:

- Live demonstration of product features and methods of operating individual product features.
- Computer-based training as either online web-based training or offline use of the training software package distributed to each trainee. Computer-based training can offer interactive self-paced lessons with frequent quizzes to ensure that the user understands the training materials.

Training must include, among other things:

- Differences between the old product, if one exists, and the new one
- The business purpose of the new product and advantages of its use
- All business operations performed with the product
- Possible issues and common errors to avoid when using the product

After developing the training materials, it is useful to conduct a pilot training program with a few specially selected users to evaluate the effectiveness of the training and adjust the materials if necessary. If the list of trainees is large, a "train-the-trainer" program may be used, where an advanced group of trained users conducts training for the other users. Otherwise, a selected group of delivery team members or well-trained business team members is assigned to conduct the training.

Deliver Training (T5-4)

In order to deliver the training, three steps must be followed:

1. Schedule each training session and send notifications to trainees
2. Assign trainees for each training session
3. Distribute training materials and conduct training sessions

With the exception of computer-based training, training sessions must be scheduled in advance to avoid interruption of existing business. All trainees and their managers must be notified and acceptance must be confirmed. The entire training schedule should be sent to the trainees, so they can request an alternate date if they cannot attend on the scheduled date. Their managers must release them from other duties during the training session and assign other personnel to perform their duties during that time. If a trainee misses part of the training, he or she does not receive training certification and is not allowed to operate the product until complete training is received in an alternate training session. The training schedule is considered to be finalized after the trainees' acceptance is confirmed and training rooms and training equipment are booked for all sessions.

Training sessions should preferably be delivered by trained members of the business team. Training sessions should not extend beyond six hours a day, with a 15-minute break every two hours. Longer sessions reduce the effectiveness of the training. If training sessions are longer than three hours, there should be two trainers, alternating every hour or two.

Project Closeout (T6)

The closeout process starts when both the Project Rollout (T4) and User Training (T5) processes are complete. The major purpose of project closeout includes the following key processes, as displayed in Figure 16.7:

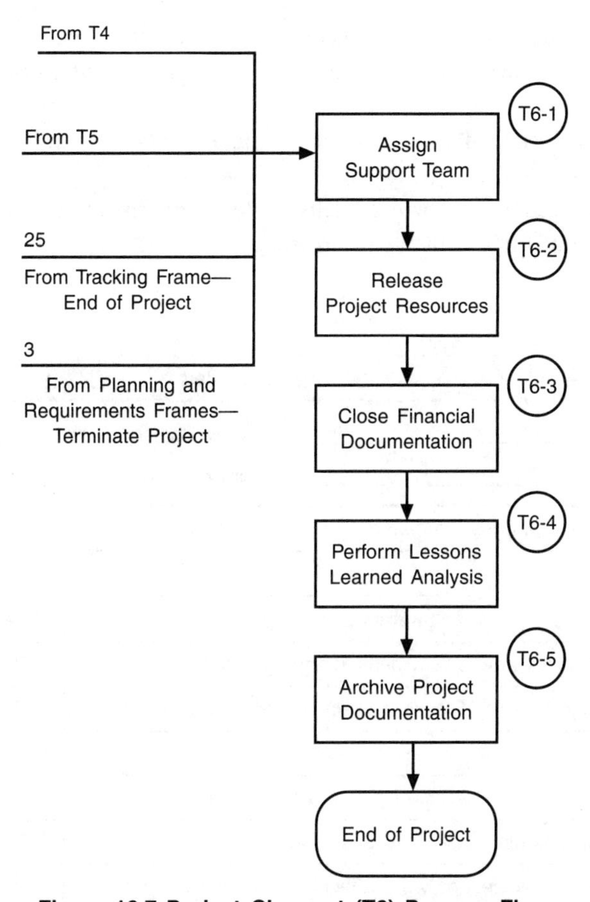

Figure 16.7 Project Closeout (T6) Process Flow

1. Assign Support Team (T6-1)
2. Release Project Resources (T6-2)
3. Close Financial Documentation (T6-3)
4. Perform Lessons Learned Analysis (T6-4)
5. Archive Project Documentation (T6-5)

Assign Support Team (T6-1)

When the project cutoff to the production or business environment is concluded in the Project Rollout (T4) process, a special assigned team must continue to exercise some control of the product for maintenance and operation support even after the project delivery team no longer exists. To enable such a team to perform its duty, the transfer of knowledge and documentation should be completed at this point.

Release Project Resources (T6-2)

Project resources, such as project staff, equipment and facilities, must be released in process T6-2 in the following sequence:

1. All delivery team staff provide feedback by completing the end-of-project review form (Table 16.6), which can provide valuable insight into the inner project issues.
2. Collect from team members all project documents and equipment they used for project work.
3. Release the remaining project delivery team members and return them to the resource pool for use in other projects.

Table 16.6 End-of-Project Review Form

End-of-Project Review				
Project Name: ________		Review Completed on: ________		
Project Manager: ________		Team Member: ________		
Note: If any of your responses to statements are FALSE or N/A, please provide explanation.				
		Response		
#	**Statement**	**True**	**False**	**N/A**
1	Your project duties were clearly defined and agreed upon with you.	☐	☐	☐
2	All your assignments and tasks were clear and unambiguous.	☐	☐	☐
3	You completed most of your assignments on time.	☐	☐	☐
4	Most predecessor tasks assigned to others were completed on time.	☐	☐	☐
5	Most of your task estimates were accurate.	☐	☐	☐
6	Most project team meetings were effective and efficient.	☐	☐	☐
7	You never had personal issues which affected your work.	☐	☐	☐
8	You were almost never required to perform other unscheduled duties or devote excessive time to participation in meetings, conferences, etc.	☐	☐	☐
9	You almost never had to rework work completed earlier.	☐	☐	☐
10	You never did any work without the project manager's knowledge.	☐	☐	☐
11	Development or project management processes never hampered your work.	☐	☐	☐
12	If you had to start the project all over, you would do everything the same way.	☐	☐	☐
13	You were never frustrated with the project.	☐	☐	☐
14	Your interaction with the project manager was reasonable and business-like.	☐	☐	☐
15	You felt that your project work was appreciated and recognized.	☐	☐	☐
Please feel free to provide your comments:				

4. Submit performance evaluations of all delivery team members to their resource owners and in some cases provide recommendations for recognition of performance.
5. Return equipment and facilities to owners.

Close Financial Documentation (T6-3)

The project manager will advise the accounting or finance department that the project is complete. Based on this, all project financial documentation, such as bills, payments, purchase orders and the project contract, should be closed and archived. At the same time, a new contract may be established for product or service maintenance. A formal letter must be sent to the client announcing contract completion. In the case of premature project termination, the reason for termination must be stated.

Perform Lessons Learned Analysis (T6-4)

A lessons learned analysis session must be conducted for all projects, whether they were successful or not, in order to avoid making the same mistakes in other projects. Major participants in the session are the project manager, the technical lead and the client.

The collective wisdom of previous project managers is a valuable resource for the organization. The opportunity to capture their wisdom for practical use should not be lost.

All events that occurred during the course of the project are analyzed and a determination is made as to whether the response applied was the correct one. This should be recorded in the project control book as it happens, instead of only at the end of the project, when memory fades or becomes distorted, key contributors have left and the pain (or pleasure) of the moment has passed.

For incorrect responses or actions taken during the course of the project, it is important to identify what should have been done instead. Along with events, project management processes should be scrutinized and their effectiveness checked. In addition, the end-of-project forms completed by the delivery team members should be thoroughly reviewed. Every problem should be discussed and documented in the following format:

1. Present the problem.
2. Present corrective actions taken to resolve the problem.
3. Review results of the problem resolution and analyze any undesirable effects that occurred.
4. Discuss what a correct course of action would have been.
5. Provide recommendation about an overall better approach to the project.
6. Review the end-of-project feedback forms and identify significant issues.

The lessons learned analysis must be documented in the project control book. The analysis should be made available to delivery management for follow-up to avoid similar mistakes in future projects—or to build on the successes of the project under analysis. It is useful to make these findings available in a searchable database so that corrective actions for many problems can be used in other projects.

Archive Project Documentation (T6-5)

All project documentation contained in the project control book should be archived for future use and for auditing purposes. This information does not have to be available for immediate retrieval; instead, a copy can be ordered, but the original stays in the archive.

Before archiving the project control book, all documentation must be made "read-only" to avoid modifications. Any electronic copy retrieved should retain the "read-only" attribute. When the copy is no longer needed, it should be discarded by the administrator.

Appendices

Project Workflow Process Hierarchy

Requirements Frame

R1 Receive Initial Project Request and Benefit Statement
This process provides the delivery team with a general idea about the project and outlines benefits expected from the project.

R2 Create Project Control Book
The project control book is a tool for keeping all project documentation in one place.

R3a and R3b Perform/Update Cost-Benefit Analysis
This process calculates whether benefits from the project justify expenses.

R4 Business Requirements Analysis
The purpose of this process is to elicit detailed project requirements and analyze them.

R4-1 Elicit Detailed Requirements
R4-2 Assign Requirements Priority
R4-3 Establish Acceptance Criteria

R5a and R5b Create/Update Traceability Matrix
The traceability matrix is a tool for tracking all changes to the project scope throughout the project life cycle.

R6 Create Business Requirements Document
This document outlines the baseline for all project requirements.

R7 Conduct Requirements Review

This process ensures the quality of business requirements and verifies correctness of the business requirements document.

R7-1 Identify Review Team
R7-2 Schedule Review and Invite Participants
R7-3 Send Materials to Participants
R7-4 Conduct Review/Take Notes
R7-5 Update Materials

R8 Approve Requirements

This process approves requirements and authorizes funds.

R8-1 Obtain Requirements Manager Sign-Off, Followed by Project Manager Sign-Off
R8-2 Obtain Lead Client Sign-Off
R8-3 Obtain Delivery Manager Sign-Off
R8-4 Obtain Senior Business Manager Sign-Off

R9 Conduct Kickoff Meeting

The kickoff meeting is the official project initiation.

R10 Review Planning Frame Milestones with Client

This process reviews the planning/high-level design and the overall project plan or milestones with the client before requesting project authorization. Plans may be changed later as a result of the review.

R11 Update Planning Frame Plan

This process updates plans or milestones with changes requested at the review before beginning the Planning/High-Level Design Frame implementation.

R12 Request Project Authorization

This process requests project approval and the project charter from the project sponsor. Receipt confirms that the sponsor agrees with the project scope, including the cost and schedule of at least the Planning/High-Level Design Frame activities.

Planning/High-Level Design Frame

P1 Risk Management Planning

The purpose of this process is to identify and document risks, control the impact of risks and to minimize the effects of threats and maximize the effects of opportunities.

P1-1 Identify Potential Risks
P1-2 Determine Probability of Occurrence
P1-3 Determine Maximum Loss Value of Each Risk
P1-4 Calculate Expected Monetary Value (EMV) of Each Risk
P1-5 Develop Risk Response Plan for Each Risk

P1-6 Balance EMV of Acceptable Project Risks
P1-7 Calculate Total Risk Costs
P1-8 Determine Severity of Each Risk
P1-9 Calculate Each Risk Rating
P1-10 Balance Ratings of Acceptable Project Risks

P2 Quality Management Planning

This process establishes quality management plans, quality checklists and the schedule of quality assurance (QA) audits and quality control (QC) reviews. Quality management controls quality throughout the project life cycle, from inception to closing. Without this process, there would be no way to determine whether or not the project implementation is successful.

P2-1 QA Audit
P2-1-1 Inform Project Manager of QA Audit and Send Checklists
P2-1-2 Answer Checklist Questions
P2-1-3 Schedule QA Audit
P2-1-4 Conduct QA Audit
P2-1-5 Eliminate Gaps
P2-1-6 Rate QA Audit
P2-1-7 Escalate Noncompliance
P2-2 QC Review
P2-2-1 Identify Review Team
P2-2-2 Schedule Review and Invite Participants
P2-2-3 Send Materials to Participants
P2-2-4 Conduct Review/Take Notes
P2-2-5 Update Materials
P2-2-6 Rate QC Review
P2-2-7 Modify Deliverable

P3 Configuration Management Planning

The purpose of this process is to control project components at any given time, preventing unauthorized changes to the completed deliverables, and to produce the schedule of configuration management (CM) reviews.

P3-1 Establish Configuration Control Board
P3-2 Develop CM Plan
P3-3 Establish CM Baseline
P3-4 Conduct CM Baseline Audit
P3-5 Approve Promotion (to stage/production libraries or warehouse)
P3-6 Escalate
P3-7 Correct Discrepancies

P4 Issue Management Planning

The purpose of this process is to identify and manage issues that come up during the project frames and to resolve issues and minimize their impact on the project throughout its life cycle. The issue management plan identifies resources responsible for each issue resolution task, resources for escalation when needed and the target dates for issue resolution. The process is triggered every time an issue comes up. In some cases, when an issue cannot be resolved without a scope change, it triggers a new scope change request.

P4-1 Identify/Modify and Document Issue
P4-2 Assign Ownership
P4-3 Propose or Modify Issue Resolution
P4-4 Escalate Issue
P4-5 Close Issue

P5 and P5a Work Breakdown Structure Design and Preliminary Project Planning

The purpose of the work breakdown structure (WBS) is to establish plans for managing the project. The WBS contains the project information (e.g., milestones, deliverables, dependencies, risks, tasks, etc.) and identifies resource requirements and training plans. The WBS utilizes inputs from processes P1 through P6, P11 and P14. Changes to the project scope will bring the project flow back to this process. See the note in Chapter 1 about the WBS, explaining why the meaning of the WBS in this book is different from the existing standard.

P6 Communication Management Planning

Communication management is required to identify stakeholders and develop project reporting and the reporting templates for different types of project communication. This process produces the schedule of communications to stakeholders.

P7 Scope Change Control

The project scope change process establishes rules for implementing scope changes, while avoiding scope leak and scope creep. Scope leak occurs when some project scope elements are missing in the end product, while scope creep occurs when additional undocumented scope elements are present in the end product or the delivered project scope differs from the documented scope. This process produces the implementation schedule for each formal scope change request (SCR). An SCR must be received before planning a scope change.

P7-1 Submit SCR
P7-2 Review SCR for Evaluation
P7-3 Notify Project Stakeholders
P7-4 Approve Budget for SCR Analysis
P7-5 Approve SCR Implementation
P7-6 Close SCR
P7-7 Wait for End of SCR Deferral

P8 Create/Update Plan Package

The project plan package is a set of the various plans, schedules, resource assignments, estimates, etc. required for project budget approval.

P9 Approve Construction Frame Plan and Budget

This process is executed at the end of the Planning Frame just before the Construction Frame starts. The final budget must have a definitive accuracy of –5% to +10%.

P10 Create Statement of Work

The purpose of this process is to describe the steps necessary to create the statement of work, which is the single most important legal document, as it lays the foundation for the project.

P11 High-Level Design/Architecture

This process starts execution at the same time as all the other planning processes (P1 through P6) and must be finished before the plan package is produced in process P8. The flow for this process is purely technical in nature and will be entirely different for different industries.

P12a and P12b Estimating and Ballpark Estimating

The purpose of estimating is to produce size, effort, cost, schedule and critical resource estimates for a project or a frame throughout its life cycle. The estimating process examines the activities necessary and the effort required to accomplish them. Process P12b produces only ballpark estimates, while process P12a produces either preliminary estimates for the entire project or definitive estimates for the project frame being planned.

- P12-1 Estimate Size
- P12-2 Estimate Effort
    - P12-2-1 Perform/Adjust Estimates
    - P12-2-2 Validate Estimates
    - P12-2-3 Add Overhead to Estimates
    - P12-2-4 Perform Alternate Method Estimates
- P12-3 Estimate Cost
- P12-4 Estimate Schedule
- P12-5 Estimate Critical Resources
- P12-6 Review Estimates
    - P12-6-1 Identify Review Team
    - P12-6-2 Schedule Review
    - P12-6-3 Conduct Review
    - P12-6-4 Record Review Notes
    - P12-6-5 Update Estimates
    - P12-6-6 Confirm Estimates

P13 Approve Closing Frame Plan and Budget

This process is executed at the end of the Construction/Tracking Frame, just before the Closing/Testing Frame starts. The final budget determined here has an estimated definitive accuracy of –5% to +10%.

P14 Outsourcing/Offshore Management Planning

The purpose of outsourcing management is selection of qualified suppliers for implementation of project components, as well as managing the relationship with them for quality deliverables and seamless integration with other components of the project.

P15 Resource Management Planning

The purpose of resource management is to obtain and manage resources working on the project.

P15-1 Plan Resources
P15-2 Acquire Resources
P15-3 Assign Resources to Tasks
P15-4 Build the Team

Construction/Tracking Frame

C1 Earned Value Analysis and Tracking

Earned value analysis is a widely used method to keep track of project work and report project financials and project health based on the earned value analysis tracking acceptance criteria.

C2 Quality Management Implementation and Tracking

Quality management implementation implements the planned quality reviews and audits. Quality management tracking compares the planned versus the actually executed quality reviews and audits in accordance with the quality management tracking acceptance criteria.

C3 Risk Management Implementation and Tracking

Risk management implementation is the execution of planned and sometimes unplanned risk assessments. Risk monitoring/tracking monitors risk occurrence in accordance with the risk acceptance criteria.

C4 Communication Management Implementation and Tracking

Communication management implementation is maintaining communication between project team members in the delivery, subcontractor and business organizations. Communication management tracking compares actual dates when meetings, orientation and training took place versus the communication plan and communication acceptance criteria.

C5 Outsourcing Management Implementation and Tracking

Tracking the outsourced parts of the project is based on status and financial reports provided by the outsourcing organization. Variances are tracked according to the acceptance criteria.

C6 Scope Change Implementation and Tracking
Scope change tracking is the tracking of actuals versus the plan in accordance with the acceptance criteria.

C7 Detailed Design and Construction
The main goal of this process is implementation of the existing project plan and its tracking according to the acceptance criteria.

C8 Troubled Project Assessment
This process determines major reasons why the project is unable to proceed as planned.

C9 Take Corrective Action
Corrective action is taken in order to eliminate reasons affecting project performance and improve it to be in line with the project baseline, quality or expected results.

Closing/Testing Frame

T1 Acceptance Test Planning
The purpose of this process is to detail the steps necessary to run the acceptance test. The process includes test planning, test scripts development and defect management planning. This is a business-owned process.

- T1-1 Analyze Business Requirements Document and Prepare Test Plan
  - T1-1-1 Analyze Business Requirements Document
  - T1-1-2 Identify Test Conditions for Each Requirement in Business Requirements Document
  - T1-1-3 Identify Test Cases for Each Requirement in Business Requirements Document
  - T1-1-4 Break Down Each Test Case for Detailed Elementary Test Steps
  - T1-1-5 Document Test Cases
- T1-2 Build, Document and Review Test Cases
- T1-3 Prepare Test Execution Checklist

T2 Test Scripts Execution
This process tests functionality of the product developed in accordance with the scenarios or scripts developed. This is a business-owned process.

- T2-1 Open Next Script
- T2-2 Execute Script
- T2-3 Approve Incident
- T2-4 Open New Incident Report
- T2-5 Document Script Execution
- T2-6 Receive Defect Resolution Notification
- T2-7 Close Incident
- T2-8 Reschedule Test

T3 Defect Resolution

This process identifies the steps necessary to fix a defect by performing root cause analysis and the steps to track defects that occur before and after release of the product into production. This is a delivery-team-owned process.

T3-1 Get Incident Notification
T3-2 Review Incident
T3-3 Clarify Information
T3-4 Assign Defect Owner
T3-5 Resolve Defect
T3-6 Track Resolution
T3-7 Escalate
T3-8 Perform Root Cause Analysis
T3-9 Notify Test Manager

T4 Project Rollout

The purpose of this process is to establish methods for project rollout and to roll out the project to the production environment or deliver the product to users.

T5 User Training

The purpose of this process is to identify training needs and establish user training.

T5-1 Obtain List of Trainees
T5-2 Assess User Needs
T5-3 Develop Training Materials
T5-4 Deliver Training

T6 Project Closeout

The purpose of this process is to close project documentation, release resources and analyze the finished project in order to identify successes and failures during project execution so as to repeat successes and avoid failures in the future.

T6-1 Assign Support Team
T6-2 Release Project Resources
T6-3 Close Financial Documentation
T6-4 Perform Lessons Learned Analysis
T6-5 Archive Project Documentation

Frames Interaction Diagram

A summary of all inputs/outputs and the interaction of frames is shown in the figure in this appendix

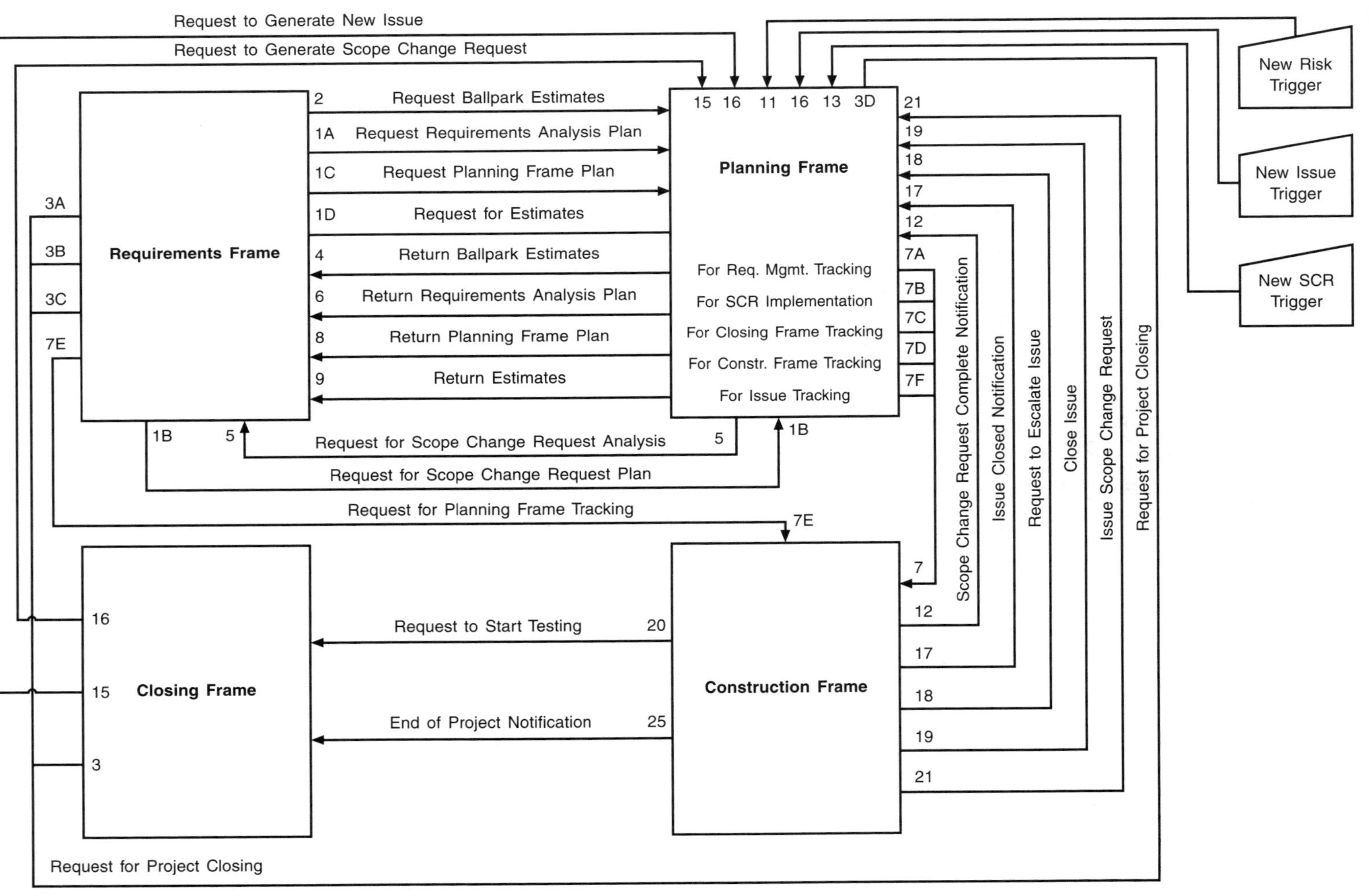

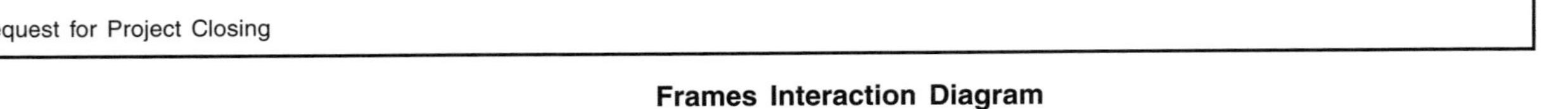
Frames Interaction Diagram

Keeping Track of Multithreaded Processes

When several threads are running at the same time, it is possible to lose track of some processes or even whole threads, since this information is not available in the project plan. In order to follow the entire process flow, it is useful to have a simple table to record which thread and which processes are being executed. A sample of multithreaded process tracking is provided in this appendix.

Let us suppose today is 11/12/2013. Earlier, during project execution, while running the Quality Management Implementation and Tracking (C2) process in thread 22, an issue came up. In order to resolve the issue, a new thread 23 was opened and the Issue Management Planning (P4) process started execution. When it became necessary to open a new scope change request (SCR) while executing the Propose or Modify Issue Resolution (P4-3) process, the issue was closed in process P4-5 and a new SCR was opened in thread 24. The SCR was planned and after SCR implementation was approved in process P7-5. Then, the scope change implementation and tracking started in thread 24.

Reviewing the above scenario, it is obvious that only thread 23 is complete. Thread 24 is still open, because the SCR cannot be closed while thread 25 is still running, implementing the scope change. Once it is complete, the process flow will return to thread 24 and process P7-6 will be executed, closing the SCR. Right after that, the flow will return to thread 22 and the Quality Management Implementation and Tracking (C2) process will complete one loop.

By looking at the sample tracking table in this appendix, it is easy to see which threads and processes are still open and need attention.

Thread Execution Tracking Table

Project Name: ______________________________

Project Manager: ______________________________

Thread #	Process ID/Name	Start	End
......	....	....	....
......	....	....	....
22		10/19/13	NA
	Quality Management Implementation and Tracking (C2)	10/19/13	NA
23		10/24/13	10/28/13
	Issue Management Planning (P4)	10/24/13	10/28/13
	Identify/Modify and Document Issue (P4-1)	10/24/13	10/24/13
	Assign Ownership (P4-2)	10/25/13	10/25/13
	Propose or Modify Issue Resolution (P4-3)	10/28/13	10/28/13
	Close Issue (P4-5)	10/28/13	10/28/13
24		10/31/13	NA
	Scope Change Control (P7)	10/31/13	NA
	Submit SCR (P7-1)	10/31/13	10/31/13
	Review SCR for Evaluation (P7-2)	11/1/13	11/3/13
	Notify Project Stakeholders (P7-3)	11/4/13	11/4/13
	Approve Budget for SCR Analysis (P7-4)	11/7/13	11/10/13
	Approve SCR Implementation (P7-5)	11/10/13	11/12/13
	Close SCR (P7-6)	NA	NA
25		11/13/13	NA
	Scope Change Implementation and Tracking (C6)	11/13/13	NA

Index

D

E

F

G

H

I

J

L

M

Q

R

S

V

W